Annals of Mathematics Studies

Number 131

THE WILLIAM H. ROEVER LECTURES IN GEOMETRY

The William H. Roever Lectures in Geometry were established in 1982 by his sons William A. and Frederick H. Roever, and members of their families, as a lasting memorial to their father, and as a continuing source of strength for the department of mathematics at Washington University, which owes so much to his long career.

After receiving a B.S. in Mechanical Engineering from Washington University in 1897, William H. Roever studied mathematics at Harvard University, where he received his Ph.D. in 1906. After two years of teaching at the Massachusetts Institute of Technology, he returned to Washington University in 1908. There he spent his entire career, serving as chairman of the Department of Mathematics and Astronomy from 1932 until his retirement in 1945.

Professor Roever published over 40 articles and several books, nearly all in his specialty, descriptive geometry. He served on the council of the American Mathematical Society and on the editorial board of the Mathematical Association of America and was a member of the mathematical societies of Italy and Germany. His rich and fruitful professional life remains an important example to his Department.

––––––––––

This monograph is an elaboration of a series of lectures delivered by William Fulton at the 1989 William H. Roever Lectures in Geometry, held on June 5–10 at Washington University, St. Louis, Missouri.

Introduction to Toric Varieties

by

William Fulton

THE WILLIAM H. ROEVER LECTURES
IN GEOMETRY

WASHINGTON UNIVERSITY
ST. LOUIS

PRINCETON UNIVERSITY PRESS

———

PRINCETON, NEW JERSEY
1993

Library of Congress Cataloging-in-Publication Data

Fulton, William.
Introduction to toric varieties / by William Fulton.
p. cm.—(Annals of mathematics studies ; no. 131)
Includes bibliographical references and index.
ISBN 0-691-03332-3—ISBN 0-691-00049-2 (pbk.)
1. Toric varieties. I. Title. II. Series
QA571.F85 1993
516.3'53—dc20 93-11045

The publisher would like to acknowledge the authors of this volume for providing the camera-ready copy from which this book was printed

Princeton University Press books are printed on acid-free paper and meet the guidelines for permanence and durability of the Committee on Production Guidelines for Book Longevity of the Council on Library Resources

Second printing, with errata sheet, 1997

http://pup.princeton.edu

Printed in the United States of America

3 5 7 9 10 8 6 4 2

William H. Roever 1874–1951

Dedicated to the memory of

Jean-Louis and Yvonne Verdier

CONTENTS

PREFACE

Algebraic geometry has developed a great deal of machinery for studying higher dimensional nonsingular and singular varieties; for example, all sorts of cohomology theories, resolution of singularities, Hodge theory, intersection theory, Riemann-Roch theorems, and vanishing theorems. There has been real progress recently toward at least a rough classification of higher dimensional varieties, particularly by Mori and his school. For all this — and for anyone learning algebraic geometry — it is important to have a good source of examples.

In introductory courses this can be done in several ways. One can study algebraic curves, where much of the story of their linear systems (line bundles, projective embeddings, etc.) can be worked out explicitly for low genus.[1] For surfaces one can work out some of the classification, and work out some of the corresponding facts for the special surfaces one finds.[2] Another approach is to study varieties that arise in "classical" projective geometry: Grassmannians, flag varieties, Veronese embeddings, scrolls, quadrics, cubic surfaces, etc.[3]

Toric varieties provide a quite different yet elementary way to see many examples and phenomena in algebraic geometry. In the general classification scheme, these varieties are very special. For example, they are all rational, and, although they may be singular, the singularities are also rational. Nevertheless, toric varieties have provided a remarkably fertile testing ground for general theories. Toric varieties correspond to objects much like the simplicial complexes studied in elementary topology, and all the basic conceptss on toric varieties — maps between them, line bundles, cycles, etc. (at least those compatible with the torus action) — correspond to simple "simplicial" notions. This makes everything much more computable and concrete than usual. For this reason, we believe it provides a good companion for an introduction to algebraic geometry (but certainly not

---

[1] The numbers refer to the notes at the back of the book.

a substitute for the study of curves, surfaces, and projective geometry!).

In addition, there are applications the other way, and interesting relations with commutative algebra and lattice points in polyhedra. The geometry of toric varieties also provides a good model for how some of the compactifications of symmetric varieties look; indeed, this was the origin of their study. Although we won't study compactifications in this book, knowing about toric varieties makes them easier to understand.

The goal of this mini-course is to develop the foundational material, with many examples, and then to concentrate on the topology, intersection theory, and Riemann-Roch problem on toric varieties. These are applied to count lattice points in polytopes, and study volumes of convex bodies. The notes conclude with Stanley's application of toric varieties to the geometry of simplicial polytopes. Relations between algebraic geometry and other subjects are emphasized, even when proofs without algebraic geometry are possible.

When this course was first planned there was no accessible text containing foundational results about toric varieties, although there was the excellent introductory survey by Danilov, as well as articles by Brylinski, Jurkiewitz, and Teissier, and more technical monographs by Demazure, Kempf-Knudsen-Mumford-Saint-Donat, Ash-Mumford-Rapoport-Tai, and Oda, where most of the results about toric varieties appeared for the first time.[4] Since then the excellent book of Oda [Oda] has appeared. This allows us to choose topics based on their suitability for an introductory course, and to present them in less than their maximum generality, since one can find complete arguments in [Oda]. Oda's book also contains a wealth of references and attributions, which frees us from attempting to give complete references or to assign credits. In no sense are we trying to survey the subject. Almost all of the material, including solutions to many of the exercises, can be found in the references. We make no claims for originality, beyond hoping that an occasional proof may be simpler than the original; and some of the intersection theory on singular toric varieties has not appeared before.

These notes were prepared in connection with the 1989 William H.

Roever Lectures in Geometry at Washington University in St. Louis. I
thank D. Wright for organizing those lectures. They are based on
courses taught at Brown and the University of Chicago. I am grateful
to C. H. Clemens, D. Cox, D. Eisenbud, N. Fakhruddin, A. Grassi, M.
Goresky, P. Hall, S. Kimura, A. Landman, R. Lazarsfeld, M. McConnell, K.
Matsuki, R. Morelli, D. Morrison, J. Pommersheim, R. Stanley, and B.
Totaro for useful suggestions in response to these lectures, courses, and
preliminary versions of these notes. Readers can also thank H. Kley
and other avid proofreaders at Chicago for finding many errors.
Particular thanks are due to R. MacPherson, J. Harris, and B.
Sturmfels with whom I have learned about toric varieties, and V.
Danilov, whose survey provided the model for these courses. The
author has been supported by the NSF.

In rewriting these notes several years later, we have not
attempted to include or survey recent work in the subject.[5] In the
notes in the back, however, we have pointed out a few results which
have appeared since 1989 and are closely related to the text. These
notes also contain references for more complete proofs, or for needed
facts from algebraic geometry. We hope this will make the text more
accessible both to those using toric varieties to learn algebraic
geometry and to those interested in the geometry and applications of
toric varieties. In addition, these notes include hints, solutions, or
references for some of the exercises.

March, 1993   William Fulton
                                               The University of Chicago

E R R A T A

I thank J. Cheah, B. Harbourne, T. Kajiwara, and I. Robertson for these corrections.

Page	For	Read
12, line 12	Gordon	Gordan
15, line 16	convex subset	convex cone
30, line 7	integrally closed	integrally closed in $\mathbb{C}[M]$
37, line 9	$\mathcal{O}^*$	$\mathcal{O}$
38, line 13	$|\Delta|$	a cone in Δ
63, line 23	abelian	abelian if Δ contains a cone of maximal dimension
64		Replace lines 13-15 with:

The group $\text{Div}_T(X)$ is torsion free since it is a subgroup of $\oplus M/M(\sigma)$. Since some $M(\sigma)$ is 0, the embedding $M \to \text{Div}_T(X)$ must split, so the cokernel $\text{Pic}(X)$ is torsion free.

67, line 26	$-a_2 x$	$a_2 x$
70, line 28	ma_1	ma_2
79, lines 19, 23, 24	mD	D

September, 1996
William Fulton

Introduction to Toric Varieties

CHAPTER 1

DEFINITIONS AND EXAMPLES

1.1 Introduction

Toric varieties as a subject came more or less independently from the work of several people, primarily in connection with the study of compactification problems.[1] This compactification description gives a simple way to say what a toric variety is: it is a normal variety X that contains a torus T as a dense open subset, together with an action $T \times X \rightarrow X$ of T on X that extends the natural action of T on itself. The torus T is the torus $\mathbb{C}^* \times \ldots \times \mathbb{C}^*$ of algebraic groups, not the torus of topology, although the latter will play a role here as well. The simplest compact example is projective space $\mathbb{P}^n$, regarded as the compactification of $\mathbb{C}^n$ as usual:

$$(\mathbb{C}^*)^n \subset \mathbb{C}^n \subset \mathbb{P}^n .$$

Similarly, any product of affine and projective spaces can be realized as a toric variety.

Besides its brevity, this definition has the virtue that it explains the original name of toric varieties as "torus embeddings." Unfortunately, this name and description may lead one to think that one would be interested in such varieties only if one has a torus one wants to compactify; indeed, one may wonder if there wouldn't have been more general interest in this subject, at least in the West, if this name had been avoided. The action of the torus on a toric variety will be important, as well as the fact that it contains the torus as a dense open orbit, but the problem with this description is that it completely ignores the relation with the simplicial geometry that makes their study so interesting. At any rate, we far prefer the name "toric varieties," which is becoming more common.

In this introductory section we give a brief definition of toric varieties as we will study them; in the following sections these notions

3

will be made more complete and precise, and the basic facts assumed
here will be proved. A toric variety will be constructed from a *lattice*
N (which is isomorphic to $\mathbb{Z}^n$ for some n), and a *fan* Δ in N,
which is a collection of "strongly convex rational polyhedral cones" σ
in the real vector space $N_{\mathbb{R}} = N \otimes_{\mathbb{Z}} \mathbb{R}$, satisfying the conditions
analogous to those for a simplicial complex: every face of a cone in Δ
is also a cone in Δ, and the intersection of two cones in Δ is a face
of each. A *strongly convex rational polyhedral cone* σ in $N_{\mathbb{R}}$ is a
cone with apex at the origin, generated by a finite number of vectors;
"rational" means that it is generated by vectors in the lattice, and
"strong" convexity that it contains no line through the origin. We often
abuse notation by calling such a cone simply a "cone in N".

　　　Let M = Hom(N,$\mathbb{Z}$) denote the dual lattice, with dual pairing
denoted $\langle\ ,\ \rangle$. If σ is a cone in N, the *dual cone* $\sigma^{\vee}$ is the set of
vectors in $M_{\mathbb{R}}$ that are nonnegative on σ. This determines a
commutative semigroup

$$S_{\sigma} = \sigma^{\vee} \cap M = \{u \in M : \langle u,v \rangle \geq 0 \text{ for all } v \in \sigma\}.$$

This semigroup is finitely generated, so its corresponding "group
algebra" $\mathbb{C}[S_{\sigma}]$ is a finitely generated commutative $\mathbb{C}$-algebra. Such
an algebra corresponds to an affine variety: set

$$U_{\sigma} = \text{Spec}(\mathbb{C}[S_{\sigma}]).$$

If τ is a face of σ, then S_{σ} is contained in S_{τ}, so $\mathbb{C}[S_{\sigma}]$ is a
subalgebra of $\mathbb{C}[S_{\tau}]$, which gives a map $U_{\tau} \to U_{\sigma}$. In fact, U_{τ} is a
principal open subset of U_{σ}: if we choose $u \in S_{\sigma}$ so that $\tau = \sigma \cap u^{\perp}$,
then $U_{\tau} = \{x \in U_{\sigma} : u(x) \neq 0\}$. With these identifications, these affine
varieties fit together to form an algebraic variety, which we denote by
$X(\Delta)$. (The "embedding" notation for this is $T_N\text{emb}(\Delta)$, but we won't
follow this convention.) Note that smaller cones correspond to smaller
open sets, which explains why the geometry in N is preferred to the
equivalent geometry in the dual space M.

　　　We turn to some simple examples. For these, the lattice N is
taken with a fixed basis $e_1, \ldots, e_n$, with $X_1, \ldots, X_n$ the elements in
$\mathbb{C}[M]$ corresponding to the dual basis. For $n \leq 3$, we usually write X,
Y, and Z for the first three of these. We first consider some affine

examples, where Δ consists of a cone σ together with all of its faces, and $X(\Delta)$ is the affine variety U_σ.

The origin $\{0\}$ is a cone, and a face of every other cone. The dual semigroup is all of M, with generators $\pm e_1^*, \ldots, \pm e_n^*$, so the corresponding group algebra is

$$\mathbb{C}[M] = \mathbb{C}[X_1, X_1^{-1}, X_2, X_2^{-1}, \ldots, X_n, X_n^{-1}],$$

which is the affine ring of the torus: $U_{\{0\}} = T = (\mathbb{C}^*)^n$. So every toric variety contains the torus as an open subset.

If σ is the cone generated by $e_1, \ldots, e_n$, then S_σ is generated by the dual basis, so

$$\mathbb{C}[S_\sigma] = \mathbb{C}[X_1, X_2, \ldots, X_n],$$

which is the affine ring of affine space: $U_\sigma = \mathbb{C}^n$.

For another example take $n = 2$, and take σ generated by e_2 and $2e_1 - e_2$.

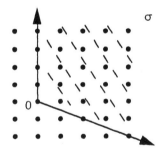
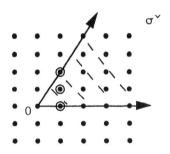

Semigroup generators for S_σ are e_1^*, $e_1^* + e_2^*$ and $e_1^* + 2e_2^*$, so

$$\mathbb{C}[S_\sigma] = \mathbb{C}[X, XY, XY^2] = \mathbb{C}[U,V,W]/(V^2 - UW).$$

Hence, U_σ is a quadric cone, i.e., a cone over a conic:

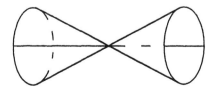

Next we look at a few basic examples which are not affine. For
$n = 1$, the only non-affine example has Δ consisting of the cones
$\mathbb{R}_{\geq 0}$, $\mathbb{R}_{\leq 0}$, and $\{0\}$, which correspond to the affine toric varieties $\mathbb{C}$,
$\mathbb{C}$, and $\mathbb{C}^*$. These three cones form a fan, and the corresponding toric
variety is constructed from the gluing:

$$\mathbb{C}[X^{-1}] \hookrightarrow \mathbb{C}[X,X^{-1}] \hookleftarrow \mathbb{C}[X]$$

$$\mathbb{C} \quad \hookleftarrow \quad \mathbb{C}^* \quad \hookrightarrow \quad \mathbb{C}$$

with the patching isomorphism given by $x \mapsto x^{-1}$ on the overlap.
This is of course the projective line $\mathbb{P}^1$.

Consider, for $n = 2$, the fan pictured:

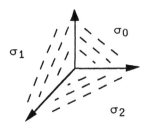

This time $U_{\sigma_0} = \mathrm{Spec}(\mathbb{C}[X,X^{-1}Y]) = \mathbb{C}^2$ and $U_{\sigma_1} = \mathrm{Spec}(\mathbb{C}[Y,XY^{-1}]) = \mathbb{C}^2$.
The resulting toric variety is the blow-up of $\mathbb{C}^2$ at the origin. To see
this, realize the blow-up as the subvariety of $\mathbb{C}^2 \times \mathbb{P}^1$ defined by the
equation $XT_1 = YT_0$, where T_0 and T_1 are homogeneous
coordinates on $\mathbb{P}^1$. This has an open cover by the two varieties U_0
and U_1 where T_0 and T_1 are nonzero, each isomorphic to $\mathbb{C}^2$; on
U_0 coordinates are X and $T_1/T_0 = X^{-1}Y$, and on U_1 coordinates are
Y and $T_0/T_1 = XY^{-1}$, which coincides with the toric construction.

Next, for $n = 2$, consider the fan

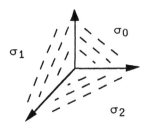

The dual cones in $M = \mathbb{Z}^2$ are

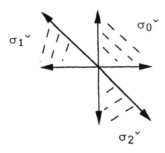

Each U_{σ_i} is isomorphic to $\mathbb{C}^2$, with coordinates (X,Y) for σ_0, $(X^{-1}, X^{-1}Y)$ for σ_1, and (Y^{-1}, XY^{-1}) for σ_2. These glue together to form the projective plane $\mathbb{P}^2$ in the usual way: if $(T_0 : T_1 : T_2)$ are the homogeneous coordinates on $\mathbb{P}^2$, $X = T_1/T_0$ and $Y = T_2/T_0$. (Note again how the geometry in N is more agreeable than that in M.)

For a more interesting example, consider a fan as drawn, where the slanting arrow passes through the point $(-1,a)$, for some positive integer a.

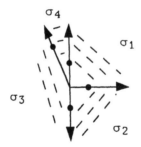

The four affine varieties are $U_{\sigma_1} = \operatorname{Spec}(\mathbb{C}[X,Y])$, $U_{\sigma_2} = \operatorname{Spec}(\mathbb{C}[X,Y^{-1}])$, $U_{\sigma_3} = \operatorname{Spec}(\mathbb{C}[X^{-1}, X^{-a}Y^{-1}])$, and $U_{\sigma_4} = \operatorname{Spec}(\mathbb{C}[X^{-1}, X^a Y])$. We have the patching

$$
\begin{array}{ccccc}
U_{\sigma_4} & (x^{-1}, x^a y) & \longleftrightarrow & (x,y) & U_{\sigma_1} \\[4pt]
& \updownarrow & & \updownarrow & \\[4pt]
U_{\sigma_3} & (x^{-1}, x^{-a} y^{-1}) & \longleftrightarrow & (x, y^{-1}) & U_{\sigma_2}
\end{array}
$$

which projects to the patching $x^{-1} \longleftrightarrow x$ of $\mathbb{C}$ with $\mathbb{C}$. The
varieties U_{σ_1} and U_{σ_2} (and U_{σ_3} and U_{σ_4}) patch together to the
variety $\mathbb{C} \times \mathbb{P}^1$, so all together we have a $\mathbb{P}^1$-bundle over $\mathbb{P}^1$. These
rational ruled surfaces are sometimes denoted $\mathbb{F}_a$, and called
Hirzebruch surfaces.

Exercise. Identify the bundle $\mathbb{F}_a \to \mathbb{P}^1$ with the bundle $\mathbb{P}(\mathcal{O}(a) \oplus \mathbb{1})$
of lines in the vector bundle that is the sum of a trivial line bundle and
the bundle $\mathcal{O}(a)$ on $\mathbb{P}^1$.[2]

Each of the four rays τ determines a curve D_τ in the surface.
Such a curve will be contained in the union of the two open sets U_σ
for the two cones σ of which τ is a face, meeting each of them in a
curve isomorphic to $\mathbb{C}$, glued together as usual to form $\mathbb{P}^1$. The
equation for $D_\tau \cap U_\sigma$ in $U_\sigma \cong \mathbb{C}^2$ is $\chi^u = 0$, where u is the
generator of S_σ that does *not* vanish on τ. For example, if τ is
the ray through e_2, the curve D_τ is defined by the equation $Y = 0$
on $U_{\sigma_1} = \text{Spec}(\mathbb{C}[X,Y])$, and $X^a Y = 0$ on $U_{\sigma_4} = \text{Spec}(\mathbb{C}[X^a Y, X^{-1}])$.

Exercise. Verify that $D_\tau \cong \mathbb{P}^1$. Show that the normal bundle to D_τ
in $\mathbb{F}_a$ is the line bundle $\mathcal{O}(-a)$, so the self-intersection number $(D \cdot D)$
is $-a$. Find the corresponding numbers for the other three rays.[3]

Beginners are encouraged to experiment before going on. See if
you can find fans to construct the following varieties as toric varieties:
$\mathbb{P}^n$, the blow-up of $\mathbb{C}^n$ at a point, $\mathbb{C} \times \mathbb{P}^1$, $\mathbb{P}^1 \times \mathbb{P}^1$, $\mathbb{C}^a \times \mathbb{P}^b$, and
$\mathbb{P}^a \times \mathbb{P}^b$. What are all the one-dimensional toric varieties? Construct
some other two-dimensional toric varieties.

1.2 Convex polyhedral cones

We include here the basic facts about convex polyhedral cones that will
be needed. These results can be found in their natural generality in
any book on convexity,[4] but the proofs in the polyhedral case are so
simple that it is nearly as easy to prove them as to quote texts. We
include proofs also because they show how to find generators of the
semigroups, which is what we need for actual computations.

Let V be the vector space $N_{\mathbb{R}}$, with dual space $V^* = M_{\mathbb{R}}$. A *convex polyhedral cone* is a set

$$\sigma = \{r_1 v_1 + \ldots + r_s v_s \in V : r_i \geq 0\}$$

generated by any finite set of vectors $v_1, \ldots, v_s$ in V. Such vectors, or sometimes the corresponding rays consisting of positive multiples of some v_i, are called *generators* for the cone σ.

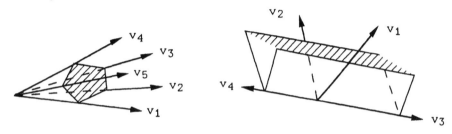

We will soon see a dual description of cones as intersections of half-spaces. The *dimension* $\dim(\sigma)$ of σ is the dimension of the linear space $\mathbb{R} \cdot \sigma = \sigma + (-\sigma)$ spanned by σ. The *dual* $\sigma^\vee$ of any set σ is the set of equations of supporting hyperplanes, i.e.,

$$\sigma^\vee = \{u \in V^* : \langle u, v \rangle \geq 0 \text{ for all } v \in \sigma\}.$$

Everything is based on the following fundamental fact from the theory of convex sets.[5]

$(*)$ If σ is a convex polyhedral cone and $v_0 \notin \sigma$, then there is some $u_0 \in \sigma^\vee$ with $\langle u_0, v_0 \rangle < 0$.

We list some consequences of $(*)$. A direct translation of $(*)$ is the *duality theorem:*

(1) $(\sigma^\vee)^\vee = \sigma$.

A *face* τ of σ is the intersection of σ with any supporting hyperplane: $\tau = \sigma \cap u^\perp = \{v \in \sigma : \langle u, v \rangle = 0\}$ for some u in $\sigma^\vee$. A cone is regarded as a face of itself, while others are called *proper* faces. Note that *any linear subspace of a cone is contained in every face of the cone.*

(2) *Any face is also a convex polyhedral cone.*

The face $\sigma \cap u^{\perp}$ is generated by those vectors v_i in a generating set for σ such that $\langle u, v_i \rangle = 0$. In particular, we see that *a cone has only finitely many faces.*

(3)   *Any intersection of faces is also a face.*

This is seen from the equation $\cap(\sigma \cap u_i^{\perp}) = \sigma \cap (\Sigma u_i)^{\perp}$ for $u_i \in \sigma^{\vee}$.

(4)   *Any face of a face is a face.*

In fact, if $\tau = \sigma \cap u^{\perp}$ and $\gamma = \tau \cap (u')^{\perp}$ for $u \in \sigma^{\vee}$ and $u' \in \tau^{\vee}$, then for large positive p, $u' + pu$ is in $\sigma^{\vee}$ and $\gamma = \sigma \cap (u'+pu)^{\perp}$.

   A *facet* is a face of codimension one.

(5)   *Any proper face is contained in some facet.*

To see this, it suffices to show that if $\tau = \sigma \cap u^{\perp}$ has codimension greater than one, it is contained in a larger face. We may assume that σ spans V (or replace V by the space spanned by σ); let W be the linear span of τ. The images $\bar{v}_i$ in V/W of the generators of σ are contained in a half-space determined by u. By moving this half-space in the sphere of half-spaces in V/W, one can find one that contains these vectors $\bar{v}_i$ but with at least one such nonzero vector in the boundary hyperplane. In other words, there is a u_0 in $\sigma^{\vee}$ so that $u_0^{\perp}$ contains τ and at least one of the vectors v_i not in W; this means that $\sigma \cap u_0^{\perp}$ is a larger face. When the codimension of τ in σ is two, so V/W is a plane, there are exactly two such supporting lines, which proves that *any face of codimension two is the intersection of exactly two facets.*

   From this we deduce by induction on the codimension:

(6)   *Any proper face is the intersection of all facets containing it.*

Indeed, if τ is any face of codimension larger than two, from (5) we can find a facet γ containing it; by induction τ is the intersection of facets in γ, and each of these is the intersection of two facets in σ, so their intersection τ is an intersection of facets.

(7)   *The topological boundary of a cone that spans V is the union of its proper faces (or facets).*

Since a face is the intersection with a supporting hyperplane, points in proper faces of σ have points arbitrarily near which are not in σ. Since σ has interior points, by looking at the line segment from a point in a face to a point in the interior, we see likewise that points in faces have points arbitrarily near which are interior points. Conversely, if v is in the boundary of the cone σ, let $w_i \to v$, $w_i \notin \sigma$. By ($*$), there are vectors $u_i \in \sigma^\vee$ with $\langle u_i, w_i \rangle < 0$. By taking the u_i in a sphere, we find a converging subsequence, so we may assume u_i has a limit u_0. Then $u_0 \in \sigma^\vee$ and v is in the face $\sigma \cap u_0^\perp$.

When σ spans V and τ is a facet of σ, there is a $u \in \sigma^\vee$, unique up to multiplication by a positive scalar, with $\tau = \sigma \cap u^\perp$. Such a vector, which we denote by u_τ, is an equation for the hyperplane spanned by τ.

(8) *If σ spans V and $\sigma \neq V$, then σ is the intersection of the half-spaces $H_\tau = \{v \in V : \langle u_\tau, v \rangle \geq 0\}$, as τ ranges over the facets of σ.*

If v were in the intersection of the half-spaces but not in σ, take any v' in the interior of σ. Let w be the last point in σ on the line segment from v' to v. Then w is in the boundary of σ, and so is in some facet τ. Then $\langle u_\tau, v' \rangle > 0$ and $\langle u_\tau, w \rangle = 0$, so $\langle u_\tau, v \rangle < 0$, a contradiction.

The proof gives a practical procedure for finding generators for the dual cone $\sigma^\vee$. For each set of $n-1$ independent vectors among the generators of σ, solve for a vector u annihilating the set; if neither u or $-u$ is nonnegative on all generators of σ it is discarded; otherwise either u or $-u$ is taken as a generator for $\sigma^\vee$; if the $n-1$ vectors are in a facet τ, this vector will be the one denoted u_τ above. From (8) we deduce the fact known as *Farkas' Theorem:*

(9) *The dual of a convex polyhedral cone is a convex polyhedral cone.*

If σ spans V, the vectors u_τ generate $\sigma^\vee$; indeed, if u in $\sigma^\vee$ were not in the cone generated by the u_τ, applying ($*$) to this cone, there is a vector v in V with $\langle u_\tau, v \rangle \geq 0$ for all facets τ and

$\langle u,v \rangle < 0$, and this contradicts (8). If σ spans a smaller linear space $W = \mathbb{R} \cdot \sigma$, then $\sigma^{\vee}$ is generated by lifts of generators of the dual cone in $W^* = V^*/W^{\perp}$, together with vectors u and $-u$ as u ranges over a basis for $W^{\perp}$.

This shows that polyhedral cones can also be given a dual definition as the intersection of half-spaces: for generators $u_1, \ldots, u_t$ of $\sigma^{\vee}$,

$$\sigma = \{v \in V : \langle u_1, v \rangle \geq 0, \ldots, \langle u_t, v \rangle \geq 0\}.$$

If we now suppose σ is *rational*, meaning that its generators can be taken from N, then $\sigma^{\vee}$ is also rational; indeed, the above procedure shows how to construct generators u_i in $\sigma^{\vee} \cap M$.

Proposition 1. (Gordon's Lemma) *If σ is a rational convex polyhedral cone, then $S_{\sigma} = \sigma^{\vee} \cap M$ is a finitely generated semigroup.*

Proof. Take $u_1, \ldots, u_s$ in $\sigma^{\vee} \cap M$ that generate $\sigma^{\vee}$ as a cone. Let $K = \{\sum t_i u_i : 0 \leq t_i \leq 1\}$. Since K is compact and M is discrete, the intersection $K \cap M$ is finite. Then $K \cap M$ generates the semigroup. Indeed, if u is in $\sigma^{\vee} \cap M$, write $u = \sum r_i u_i$, $r_i \geq 0$, so $r_i = m_i + t_i$ with m_i a nonnegative integer and $0 \leq t_i \leq 1$. Then $u = \sum m_i u_i + u'$, with each u_i and $u' = \sum t_i u_i$ in $K \cap M$.

It is often necessary to find a point in the *relative interior* of a cone σ, i.e., in the topological interior of σ in the space $\mathbb{R} \cdot \sigma$ spanned by σ. This is achieved by taking any positive combination of $\dim(\sigma)$ linearly independent vectors among the generators of σ. In particular, if σ is rational, we can find such points in the lattice.

(10) *If τ is a face of σ, then $\sigma^{\vee} \cap \tau^{\perp}$ is a face of $\sigma^{\vee}$, with $\dim(\tau) + \dim(\sigma^{\vee} \cap \tau^{\perp}) = n = \dim(V)$. This sets up a one-to-one order-reversing correspondence between the faces of σ and the faces of $\sigma^{\vee}$. The smallest face of σ is $\sigma \cap (-\sigma)$.*

To see this, note first that the faces of $\sigma^{\vee}$ are exactly the cones $\sigma^{\vee} \cap v^{\perp}$ for $v \in \sigma = (\sigma^{\vee})^{\vee}$. If τ is the cone containing v in its relative interior, then $\sigma^{\vee} \cap v^{\perp} = \sigma^{\vee} \cap (\tau^{\vee} \cap v^{\perp}) = \sigma^{\vee} \cap \tau^{\perp}$, so every face of $\sigma^{\vee}$ has the asserted form. The map $\tau \mapsto \tau^* = \sigma^{\vee} \cap \tau^{\perp}$ is clearly order-reversing, and from the obvious inclusion $\tau \subset (\tau^*)^*$ it follows formally that $\tau^* = ((\tau^*)^*)^*$, and hence that the map is one-to-one and onto. It follows from this that the smallest face of σ is $(\sigma^\vee)^\vee \cap (\sigma^\vee)^\perp = (\sigma^\vee)^\perp = \sigma \cap (-\sigma)$. In particular, we see that $\dim(\sigma \cap (-\sigma)) + \dim(\sigma^\vee) = n$. The corresponding equation for a general face τ can be deduced by putting τ in a maximal chain of faces of σ, and comparing with the dual chain of faces in $\sigma^\vee$.

(11) *If* $u \in \sigma^\vee$, *and* $\tau = \sigma \cap u^\perp$, *then* $\tau^\vee = \sigma^\vee + \mathbb{R}_{\geq 0} \cdot (-u)$.

Since both sides of this equation are convex polyhedral cones, it is enough to show that their duals are equal. The dual of the left side is τ, and the dual of the right is $\sigma \cap (-u)^\vee = \sigma \cap u^\perp$, as required.

Proposition 2. *Let* σ *be a rational convex polyhedral cone, and let* u *be in* $S_\sigma = \sigma^\vee \cap M$. *Then* $\tau = \sigma \cap u^\perp$ *is a rational convex polyhedral cone. All faces of* σ *have this form, and*

$$S_\tau = S_\sigma + \mathbb{Z}_{\geq 0} \cdot (-u) .$$

Proof. If τ is a face, then $\tau = \sigma \cap u^\perp$ for any u in the relative interior of $\sigma^\vee \cap \tau^\perp$, and u can be taken in M since $\sigma^\vee \cap \tau^\perp$ is rational. Given $w \in S_\tau$, then $w + p \cdot u$ is in $\sigma^\vee$ for large positive p, and taking p to be an integer shows that w is in $S_\sigma + \mathbb{Z}_{\geq 0} \cdot (-u)$.

Finally, we need the following strengthening of (*), known as a *Separation Lemma*, that separates convex sets by a hyperplane:

(12) *If* σ *and* σ' *are convex polyhedral cones whose intersection* τ *is a face of each, then there is a* u *in* $\sigma^\vee \cap (-\sigma')^\vee$ *with*

$$\tau = \sigma \cap u^\perp = \sigma' \cap u^\perp .$$

This is proved by looking at the cone $\gamma = \sigma - \sigma' = \sigma + (-\sigma')$. We know that for any u in the relative interior of $\gamma^\vee$, $\gamma \cap u^\perp$ is the smallest face of γ:

$$\gamma \cap u^\perp = \gamma \cap (-\gamma) = (\sigma - \sigma') \cap (\sigma' - \sigma) .$$

The claim is that this u works. Since σ is contained in γ, u is in $\sigma^\vee$, and since τ is contained in $\gamma \cap (-\gamma)$, τ is contained in $\sigma \cap u^\perp$. Conversely, if $v \in \sigma \cap u^\perp$, then v is in $\sigma' - \sigma$, so there is an equation $v = w' - w$, $w' \in \sigma'$, $w \in \sigma$. Then $v + w$ is in the intersection τ of

σ and σ', and the sum of two elements of a cone can be in a face only if the summands are in the face, so v is in τ. This shows that $\sigma \cap u^\perp = \tau$, and the same argument for $-u$ shows that $\sigma' \cap u^\perp = \tau$.

Proposition 3. *If σ and σ' are rational convex polyhedral cones whose intersection τ is a face of each, then*

$$S_\tau = S_\sigma + S_{\sigma'} .$$

Proof. One inclusion $S_\tau \supset S_\sigma + S_{\sigma'}$ is obvious. For the other inclusion, by the proof of (12) we can take u in $\sigma^\vee \cap (-\sigma')^\vee \cap M$ so that $\tau = \sigma \cap u^\perp = \sigma' \cap u^\perp$. By Proposition 2 and the fact that $-u$ is in $S_{\sigma'}$, we have $S_\tau \subset S_\sigma + \mathbb{Z}_{\geq 0} \cdot (-u) \subset S_\sigma + S_{\sigma'}$, as required.

(13) *For a convex polyhedral cone σ, the following conditions are equivalent:*

(i) $\sigma \cap (-\sigma) = \{0\}$;

(ii) σ *contains no nonzero linear subspace;*

(iii)   *there is a u in $\sigma^\vee$ with $\sigma \cap u^\perp = \{0\}$;*

(iv) $\sigma^\vee$ *spans V^*.*

The first two are equivalent since $\sigma \cap (-\sigma)$ is the largest subspace in σ; the second two are equivalent since $\sigma \cap (-\sigma)$ is the smallest face of σ. The first and last are equivalent since $\dim(\sigma \cap (-\sigma)) + \dim(\sigma^\vee) = n$.

A cone is called *strongly convex* if it satisfies the conditions of (13). Any cone is generated by some minimal set of generators. If the cone is strongly convex, then the rays generated by a minimal set of generators are exactly the one-dimensional faces of σ (as seen by applying (*) to any generator that is not in the cone generated by the others); in particular, these minimal generators are unique up to multiplication by positive scalars.

Exercise. If τ is a face of σ, with $W = \mathbb{R} \cdot \tau$, show that $\overline{\sigma} = (\sigma + W)/W$ is a convex polyhedral cone in V/W (rational if σ is rational), and the faces of $\overline{\sigma}$ are exactly the cones of the form $\overline{\gamma} = (\gamma + W)/W$ as γ ranges over the cones of σ that contain τ.

Exercise. For a cone σ and $v \in \sigma$, show that the following are equivalent: (i) v is in the relative interior of σ; (ii) $\langle u, v \rangle > 0$

for all u in $\sigma^\vee \setminus \sigma^\perp$; (iii) $\sigma^\vee \cap v^\perp = \sigma^\perp$; (iv) $\sigma + \mathbb{R}_{\geq 0} \cdot (-v) = \mathbb{R} \cdot \sigma$; (v) for all $x \in \sigma$ there is a positive number p and a y in σ with $p \cdot v = x + y$. [6]

Exercise. If τ is a face of a cone σ, show that the sum of two vectors in σ can be in τ only if both of the summands are in τ. Show conversely that any convex subset of a cone σ satisfying this condition is a face.

Since we are mainly concerned with these cones, we will often say "σ is a cone in N" to mean that σ is a strongly convex rational polyhedral cone in $N_\mathbb{R}$. We will sometimes write "$\tau < \sigma$" or "$\sigma > \tau$" to mean that τ is a face of σ. A cone is called *simplicial*, or a *simplex*, if it is generated by linearly independent generators.

Exercise. If σ spans $N_\mathbb{R}$, must σ and $\sigma^\vee$ have the same minimal number of generators? [7]

1.3   Affine toric varieties

When σ is a strongly convex rational polyhedral cone, we have seen that $S_\sigma = \sigma^\vee \cap M$ is a finitely generated semigroup. Any additive semigroup S determines a "group ring" $\mathbb{C}[S]$, which is a commutative $\mathbb{C}$-algebra. As a complex vector space it has a basis χ^u, as u varies over S, with multiplication determined by the addition in S:

$$\chi^u \cdot \chi^{u'} = \chi^{u+u'}.$$

The unit 1 is χ^0. Generators $\{u_i\}$ for the semigroup S determine generators $\{\chi^{u_i}\}$ for the $\mathbb{C}$-algebra $\mathbb{C}[S]$.

Any finitely generated commutative $\mathbb{C}$-algebra A determines a complex affine variety, which we denote by $\mathrm{Spec}(A)$. We review this construction and its related notation.[8] If generators of A are chosen, this presents A as $\mathbb{C}[X_1, \ldots, X_m]/I$, where I is an ideal; then $\mathrm{Spec}(A)$ can be identified with the subvariety $V(I)$ of affine space $\mathbb{C}^m$ of common zeros of the polynomials in I, but as usual for modern mathematicians, it is convenient to use descriptions that are independent of coordinates. In our applications, A will be a domain, so Spec(A) will be an irreducible variety. Although Spec(A) officially includes all prime ideals of A (corresponding to subvarieties of V(I)), when we speak of a *point* of Spec(A) we will mean an ordinary closed point, corresponding to a maximal ideal, unless we specify otherwise. These closed points are denoted Specm(A). Any homomorphism $A \to B$ of $\mathbb{C}$-algebras determines a morphism Spec(B) → Spec(A) of varieties. In particular, closed points correspond to $\mathbb{C}$-algebra homomorphisms from A to $\mathbb{C}$. If X = Spec(A), for each nonzero element $f \in A$ the principal open subset

$$X_f = \text{Spec}(A_f) \subset X = \text{Spec}(A)$$

corresponds to the localization homomorphism $A \mapsto A_f$.

  For A = $\mathbb{C}[S]$ constructed from a semigroup, the points are easy to describe: they correspond to homomorphisms of semigroups from S to $\mathbb{C}$, where $\mathbb{C} = \mathbb{C}^* \cup \{0\}$ is regarded as an abelian semigroup via multiplication:

$$\text{Specm}(\mathbb{C}[S]) = \text{Hom}_{sg}(S, \mathbb{C}).$$

For a semigroup homomorphism x from S to $\mathbb{C}$ and u in S, the value of the corresponding function χ^u at the corresponding point of Specm($\mathbb{C}[S]$) is the image of u by the map x: $\chi^u(x) = x(u)$.

  When $S = S_\sigma$ arises from a strongly convex rational polyhedral cone, we set $A_\sigma = \mathbb{C}[S_\sigma]$, and

$$U_\sigma = \text{Spec}(\mathbb{C}[S_\sigma]) = \text{Spec}(A_\sigma),$$

the corresponding *affine toric variety*. All of these semigroups will be sub-semigroups of the group $M = S_{(0)}$. If $e_1, \ldots, e_n$ is a basis for N, and $e_1^*, \ldots, e_n^*$ is the dual basis of M, write

$$X_i = \chi^{e_i^*} \in \mathbb{C}[M].$$

As a semigroup, M has generators $\pm e_1^*, \ldots, \pm e_n^*$, so

$$\mathbb{C}[M] = \mathbb{C}[X_1, X_1^{-1}, X_2, X_2^{-1}, \ldots, X_n, X_n^{-1}]$$

$$= \mathbb{C}[X_1, \ldots, X_n]_{X_1 \cdot \ldots \cdot X_n},$$

which is the ring of *Laurent polynomials* in n variables. So

$$U_{\{0\}} = \text{Spec}(\mathbb{C}[M]) \cong \mathbb{C}^* \times \ldots \times \mathbb{C}^* = (\mathbb{C}^*)^n$$

is an affine algebraic *torus*. All of our semigroups S will be sub-semigroups of a lattice M, so $\mathbb{C}[S]$ will be a subalgebra of $\mathbb{C}[M]$; in particular, $\mathbb{C}[S]$ will be a domain. When a basis for M is chosen as above, we usually write elements of $\mathbb{C}[S]$ as Laurent polynomials in the corresponding variables X_i. Note that all of these algebras are generated by *monomials* in the variables X_i.

The torus $T = T_N$ corresponding to M or N can be written intrinsically:

$$T_N = \text{Spec}(\mathbb{C}[M]) = \text{Hom}(M, \mathbb{C}^*) = N \otimes_{\mathbb{Z}} \mathbb{C}^*.$$

For a basic example, let σ be the cone with generators $e_1, \ldots, e_k$ for some k, $1 \le k \le n$. Then

$$S_\sigma = \mathbb{Z}_{\ge 0} \cdot e_1^* + \mathbb{Z}_{\ge 0} \cdot e_2^* + \ldots + \mathbb{Z}_{\ge 0} \cdot e_k^* + \mathbb{Z} \cdot e_{k+1}^* + \ldots + \mathbb{Z} \cdot e_n^*.$$

Hence $A_\sigma = \mathbb{C}[X_1, X_2, \ldots, X_k, X_{k+1}, X_{k+1}^{-1}, \ldots, X_n, X_n^{-1}]$, and

$$U_\sigma = \mathbb{C} \times \ldots \times \mathbb{C} \times \mathbb{C}^* \times \ldots \times \mathbb{C}^* = \mathbb{C}^k \times (\mathbb{C}^*)^{n-k}.$$

It follows from this example that if σ is generated by k elements that can be completed to a basis for N, then U_σ is a product of affine k-space and an $(n-k)$-dimensional torus. In particular, such affine toric varieties are nonsingular.

Next we look at a singular example. Let N be a lattice of rank 3, and let σ be the cone generated by four vectors v_1, v_2, v_3, and v_4 that generate N and satisfy $v_1 + v_3 = v_2 + v_4$. The variety U_σ is a "cone over a quadric surface", a variety met frequently when singularities are studied. If we take $N = \mathbb{Z}^3$ and $v_i = e_i$ for $i = 1, 2, 3$, so $v_4 = e_1 + e_3 - e_2$,

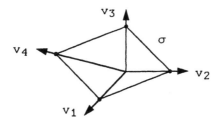

then S_σ is generated by e_1^*, e_3^*, $e_1^* + e_2^*$, and $e_2^* + e_3^*$, so

$$A_\sigma = \mathbb{C}[X_1, X_3, X_1 X_2, X_2 X_3] = \mathbb{C}[W,X,Y,Z]/(WZ - XY).$$

Therefore U_σ is the hypersurface defined by $WZ = XY$ in $\mathbb{C}^4$.

A homomorphism of semigroups $S \to S'$ determines a homomorphism $\mathbb{C}[S] \to \mathbb{C}[S']$ of algebras, hence a morphism $\mathrm{Spec}(\mathbb{C}[S']) \to \mathrm{Spec}(\mathbb{C}[S])$ of affine varieties. In particular, if τ is contained in σ, then S_σ is a sub-semigroup of S_τ, corresponding to a morphism $U_\tau \to U_\sigma$. For example, the torus $T_N = U_{\{0\}}$ maps to all of the affine toric varieties U_σ that come from cones σ in N.

Lemma. *If τ is a face of σ, then the map $U_\tau \to U_\sigma$ embeds U_τ as a principal open subset of U_σ.*

Proof. By Proposition 2 in §1.2, there is a $u \in S_\sigma$ with $\tau = \sigma \cap u^\perp$ and

$$S_\tau = S_\sigma + \mathbb{Z}_{\geq 0} \cdot (-u).$$

This implies immediately that each basis element for $\mathbb{C}[S_\tau]$ can be written in the form $\chi^{w - pu} = \chi^w / (\chi^u)^p$ for $w \in S_\sigma$. Hence

$$A_\tau = (A_\sigma)_{\chi^u},$$

which is the algebraic version of the required assertion.

Exercise. Show that if $\tau \subset \sigma$ and the mapping $U_\tau \to U_\sigma$ is an open embedding, then τ must be a face of σ. [9]

More generally, if $\varphi\colon N' \to N$ is a homomorphism of lattices such that $\varphi_\mathbb{R}$ maps a (rational strongly convex polyhedral) cone σ' in N' into a cone σ in N, then the dual $\varphi^\vee\colon M \to M'$ maps S_σ to $S_{\sigma'}$, determining a homomorphism $A_\sigma \to A_{\sigma'}$, and hence a morphism $U_{\sigma'} \to U_\sigma$.

Exercise. Show that if $S \subset S' \subset M$ are sub-semigroups, the corresponding map $\mathrm{Spec}(\mathbb{C}[S']) \to \mathrm{Spec}(\mathbb{C}[S])$ is birational if and only if S and S' generate the same subgroup of M. [10]

The semigroups S_σ arising from cones are special in several respects. First, it follows from the definition that S_σ is *saturated*, i.e., if $p \cdot u$ is in S_σ for some positive integer p, then u is in S_σ. In addition, the fact that σ is strongly convex implies that $\sigma^\vee$ spans $M_\mathbb{R}$, so S_σ generates M as a group, i.e.,

$$M = S_\sigma + (-S_\sigma) .$$

Exercise. Show conversely that any finitely generated sub-semigroup of M that generates M as a group and is saturated has the form $\sigma^\vee \cap M$ for a unique strongly convex rational polyhedral cone σ in N.

The following exercise shows that affine toric varieties are defined by *monomial equations.*

Exercise. If S_σ is generated by $u_1, \ldots, u_t$, so

$$A_\sigma = \mathbb{C}[\chi^{u_1}, \ldots, \chi^{u_t}] = \mathbb{C}[Y_1, \ldots, Y_t]/I ,$$

show that I is generated by polynomials of the form

$$Y_1{}^{a_1} \cdot Y_2{}^{a_2} \cdot \ldots \cdot Y_t{}^{a_t} - Y_1{}^{b_1} \cdot Y_2{}^{b_2} \cdot \ldots \cdot Y_t{}^{b_t} ,$$

where $a_1, \ldots, a_t, b_1, \ldots, b_t$ are nonnegative integers satisfying the equation

$$a_1 u_1 + \ldots + a_t u_t = b_1 u_1 + \ldots + b_t u_t . \tag{11}$$

If σ is a cone in N, the torus T_N acts on U_σ,

$$T_N \times U_\sigma \rightarrow U_\sigma ,$$

as follows. A point $t \in T_N$ can be identified with a map $M \rightarrow \mathbb{C}^*$ of groups, and a point $x \in U_\sigma$ with a map $S_\sigma \rightarrow \mathbb{C}$ of semigroups; the product $t \cdot x$ is the map of semigroups $S_\sigma \rightarrow \mathbb{C}$ given by

$$u \mapsto t(u) x(u) .$$

The dual map on algebras, $\mathbb{C}[S_\sigma] \rightarrow \mathbb{C}[S_\sigma] \otimes \mathbb{C}[M]$, is given by mapping χ^u to $\chi^u \otimes \chi^u$ for $u \in S_\sigma$. When $\sigma = \{0\}$, this is the usual product in the algebraic group T_N. These maps are compatible with inclusions of open subsets corresponding to faces of σ. In particular, they extend the action of T_N on itself.

Exercise. If σ is a cone in N and σ' is a cone in N', show that $\sigma \times \sigma'$ is a cone in $N \oplus N'$, and construct a canonical isomorphism

$$U_{\sigma \times \sigma'} = U_\sigma \times U_{\sigma'} .$$

Except for the following exercises, we will not be concerned with varieties arising from more general semigroups, although these are of considerable interest to algebraists.[12]

Exercise. (a) Let $S \subset \mathbb{Z}$ be the sub-semigroup generated by 2 and 3. Then

$$\mathbb{C}[S] = \mathbb{C}[X^2, X^3] = \mathbb{C}[Y, Z]/(Z^2 - Y^3) ,$$

so $\mathrm{Spec}(\mathbb{C}[S])$ is a rational curve with a cusp.
(b) Find $\mathbb{C}[S]$ when S is the sub-semigroup of $\mathbb{Z}_{\geq 0}$ generated by a pair of relatively prime positive integers.
(c) Find corresponding generators and relations for S generated by 3, 5, and 7. [13]

Exercise. If $\sigma \subset \mathbb{R}^2$ is the cone generated by e_2 and $\lambda \cdot e_1 - e_2$, with λ an irrational positive number, show that $\sigma^\vee \cap (\mathbb{Z}^2)$ is not finitely generated, and that $\mathbb{C}[\sigma^\vee \cap (\mathbb{Z}^2)]$ is not noetherian.

1.4   Fans and toric varieties

By a *fan* Δ *in* N is meant a set of rational strongly convex polyhedral cones σ in $N_\mathbb{R}$ such that

(1)   *Each face of a cone in* Δ *is also a cone in* Δ;
(2)   *The intersection of two cones in* Δ *is a face of each.*

For simplicity here we assume unless otherwise stated that fans are *finite*, i.e., that they consist of a finite number of cones. This will assure that our toric varieties are of finite type, not just locally of finite type. From now on a *cone in* N, or a *cone*, will be assumed to be a rational strongly convex polyhedral cone, unless otherwise stated.
From a fan Δ the *toric variety* $X(\Delta)$ is constructed by taking the disjoint union of the affine toric varieties U_σ, one for each σ in Δ, and gluing as follows: for cones σ and τ, the intersection $\sigma \cap \tau$ is a face of each, so $U_{\sigma \cap \tau}$ is identified as a principal open subvariety of

U_σ and of U_τ; glue U_σ to U_τ by this identification on these open subvarieties. The fact that these identifications are compatible comes immediately from the order-preserving nature of the correspondence from cones to affine varieties. The fact that the resulting complex variety is Hausdorff (or the algebraic variety is separated — or the resulting "prescheme" is a "scheme") comes from the following lemma.[14]

Lemma. *If* σ *and* τ *are cones that intersect in a common face, then the diagonal map* $U_{\sigma\cap\tau} \to U_\sigma \times U_\tau$ *is a closed embedding.*

Proof. This is equivalent to the assertion that the natural mapping $A_\sigma \otimes A_\tau \to A_{\sigma\cap\tau}$ is surjective. And this follows from the fact that $S_{\sigma\cap\tau} = S_\sigma + S_\tau$, as we saw in Proposition 3 of §1.2, which was (appropriately!) a consequence of the Separation Lemma for cones.

In particular, for any two cones σ and τ of Δ, we have the identity $U_\sigma \cap U_\tau = U_{\sigma\cap\tau}$. If σ is a cone in N, and Δ consists of σ together with all of its faces, then Δ is a fan and $X(\Delta)$ is the affine toric variety U_σ. We will see later that these are the only toric varieties that are affine.

Let us work out a few simple examples. In dimension one, with $N = \mathbb{Z}$, the only possible cones are the right or left half-lines and the origin. We have seen all possible fans, which give $\mathbb{C}$, $\mathbb{C}^*$, and $\mathbb{P}^1$.

For some simple two-dimensional examples, we draw some fans in $N = \mathbb{Z}^2$. For the fan

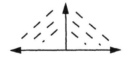

take two copies of $\mathbb{C}^2$, corresponding to the algebras $\mathbb{C}[X,Y]$ and $\mathbb{C}[X^{-1},Y]$; gluing gives $\mathbb{P}^1 \times \mathbb{C}$. Similarly, for the fan

take four copies of $\mathbb{C}^2$, and glue to get $\mathbb{P}^1 \times \mathbb{P}^1$. The preceding two examples are special cases of a general fact:

Exercise. If Δ is a fan in N, and Δ' is a fan in N', show that the set of products $\sigma \times \sigma'$, $\sigma \in \Delta$, $\sigma' \in \Delta'$, forms a fan $\Delta \times \Delta'$ in $N \oplus N'$, and show that

$$X(\Delta \times \Delta') = X(\Delta) \times X(\Delta').$$

The generalization of the construction of $\mathbb{P}^2$ is:

Exercise. Suppose vectors $v_0, v_1, \ldots, v_n$ generate a lattice N of rank n, with $v_0 + v_1 + \ldots + v_n = 0$. Let Δ be the fan whose cones are generated by any proper subsets of the vectors $v_0, \ldots, v_n$. Construct an isomorphism of $X(\Delta)$ with projective n-space $\mathbb{P}^n$.

The vectors $v_1, \ldots, v_n$ in the preceding exercise can be taken to be the standard basis $e_1, \ldots, e_n$ for $N = \mathbb{Z}^n$, with $v_0 = -e_1 - \ldots - e_n$. This corresponds to constructing $\mathbb{P}^n$ as the closure of $\mathbb{C}^n$. A more symmetric description of $\mathbb{P}^n$ can be given by taking N to be the lattice $\mathbb{Z}^{n+1}/\mathbb{Z}\cdot(1,1,\ldots,1)$, with v_i the image of the ith basic vector, $0 \le i \le n$. With this description, $T_N = (\mathbb{C}^*)^{n+1}/\mathbb{C}^*$ is embedded naturally in $\mathbb{C}^{n+1} \smallsetminus \{0\}/\mathbb{C}^* = \mathbb{P}^n$.

Exercise. Find the toric varieties corresponding to the following three fans in $\mathbb{Z}^2$:

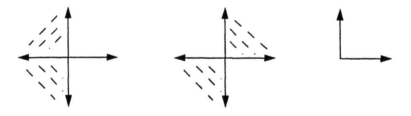

Suppose $\varphi: N' \to N$ is a homomorphism of lattices, and Δ is a fan in N, Δ' a fan in N', satisfying the following condition:

*for each cone σ' in Δ', there is some
cone σ in Δ such that $\varphi(\sigma') \subset \sigma$.*

As we saw in the preceding section, this determines a morphism

$U_{\sigma'} \to U_{\sigma} \subset X(\Delta)$. The morphism from $U_{\sigma'}$ to $X(\Delta)$ is easily seen to be independent of choice of σ, and these morphisms $U_{\sigma'} \to X(\Delta)$ patch together to give a morphism

$$\varphi_*: X(\Delta') \to X(\Delta).$$

At the end of §1.1 we constructed the Hirzebruch surface $\mathbb{F}_a$ as a toric variety. The projection $\mathbb{F}_a \to \mathbb{P}^1$ is determined by the projection $\mathbb{Z}^2 \to \mathbb{Z}$ taking (x,y) to x. (Note that the second projection does not satisfy the required condition, if $a \neq 0$.)

Exercise. Show that for any integer m, the map $\mathbb{Z} \to \mathbb{Z}^2$ given by $z \mapsto (z, mz)$ determines a section $\mathbb{P}^1 \to \mathbb{F}_a$ of this projection.

The actions of the torus T_N on the varieties U_{σ} described in the preceding section are compatible with the patching isomorphisms, giving an action of T_N on $X(\Delta)$. This extends the product in T_N:

$$\begin{array}{ccc}
T_N \times X(\Delta) & \to & X(\Delta) \\
\| \quad \updownarrow & & \updownarrow \\
T_N \times T_N & \to & T_N
\end{array}$$

The converse is also true: any (separated, normal) variety X containing a torus T_N as a dense open subvariety, with compatible action as above, can be realized as a toric variety $X(\Delta)$ for a unique fan Δ in N. We will have no use for this, so we don't discuss the proof.[15]

Although we will deal with complex toric varieties in these lectures, the interested reader should have no difficulty carrying out the same constructions over any other base field (or ground ring).

1.5   Toric varieties from polytopes

A *convex polytope* K in a finite dimensional vector space E is the convex hull of a finite set of points. A (proper) *face* F of K is the intersection with a supporting affine hyperplane, i.e.,

$$F = \{v \in K : \langle u, v \rangle = r\},$$

where $u \in E^*$ is a function with $\langle u,v \rangle \geq r$ for all v in K; K is usually included as an improper face. We assume for simplicity that K is n-dimensional, and that K contains the origin in its interior. A *facet* of K is a face of dimension n-1. The results of §1.2 can be used to deduce the corresponding basic facts about faces of convex polytopes. For this, let σ be the cone over $K \times 1$ in the vector space $E \times \mathbb{R}$. The faces of σ are easily seen to be exactly the cones over the faces of K (with the cone $\{0\}$ corresponding to the empty face of K); from this it follows that the faces satisfy the analogues of properties (2)-(7) of §1.2.

As for the duality theory of polytopes, the *polar set* (or *polar*) of K is defined to be the set

$$K^o = \{u \in E^* : \langle u,v \rangle \geq -1 \text{ for all } v \in K\}.$$

(Often $\{u \in E^* : \langle u,v \rangle \leq 1 \ \forall \ v \in K\} = -K^o$ is taken to be the polar set, but this does not change the results.) For example, the polar of the octahedron in $\mathbb{R}^3$ with vertices at the points $(\pm 1,0,0)$, $(0,\pm 1,0)$, and $(0,0,\pm 1)$ is the cube with vertices $(\pm 1,\pm 1,\pm 1)$.

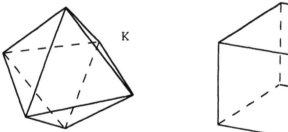
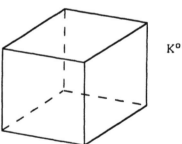

Proposition. *The polar set* K^o *is a convex polytope, and* K *is the polar of* K^o. *If* F *is a face of* K, *then*

$$F^* = \{u \in K^o : \langle u,v \rangle = -1 \ \forall \ v \in F\}$$

is a face of K^o, *and the correspondence* $F \mapsto F^*$ *is a one-to-one, order-reversing correspondence between the faces of* K *and the faces of* K^o, *with* $\dim(F) + \dim(F^*) = \dim(E) - 1$. *If* K *is rational, i.e., its vertices lie in a lattice in* E, *then* K^o *is also rational, with its vertices in the dual lattice.*

Proof. With σ the cone over $K \times 1$, the dual cone $\sigma^\vee$ consists of those $u \times r$ in $E^* \times \mathbb{R}$ such that $\langle u, v \rangle + r \geq 0$ for all v in K. It follows that $\sigma^{\vee}$ is the cone over $K^{\circ} \times 1$ in $E^* \times \mathbb{R}$. The assertions of the proposition are now easy consequences of the results in §1.2 for cones. For example, the duality $(K^{\circ})^{\circ} = K$ follows from the duality $(\sigma^{\vee})^{\vee} = \sigma$. For a face F of K, if τ is the cone over $F \times 1$, then the dual $\sigma^{\vee} \cap \tau^{\perp}$ is the cone over $F^* \times 1$, from which the duality between faces of K and K° follows.

Exercise. Let K be a *convex polyhedron* in E, i.e.,

$$K = \{ v \in E : \langle u_1, v \rangle \geq -a_1, \ldots, \langle u_r, v \rangle \geq -a_r \}$$

for some $u_1, \ldots, u_r$ in E^* and real numbers $a_1, \ldots, a_r$. Show that K is bounded if and only if K is the convex hull of a finite set.[16]

A rational convex polytope K in $N_{\mathbb{R}}$ determines a fan Δ whose cones are the cones over the proper faces of K. Since we assume that K contains the origin in its interior, the union of the cones in Δ will be all of $N_{\mathbb{R}}$. All of the fans we have seen so far whose cones cover $N_{\mathbb{R}}$ have this form.

More generally, if K' is a subdivision of the boundary of K, i.e., K' is a collection of convex polytopes whose union is the boundary of K, and the intersection of any two polytopes in K' is a polytope in K', then the cones over the polytopes in K' form a fan. Here are some examples of such K':

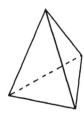 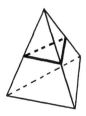 

Note that the second can be "pushed out", so that the cone over it is the cone over a convex polytope, but the third cannot.

There are many fans, however, that do not come from *any* convex polytope, however subdivided. To see one, start with the fan over the faces of the cube with vertices at $(\pm 1, \pm 1, \pm 1)$ in $\mathbb{Z}^3$. Let Δ be the fan with cones spanned by the same sets of generators except

that the vertex $(1,1,1)$ is replaced by $(1,2,3)$. It is impossible to find
eight points, one on each of the eight positive rays through the
vertices, such that for each of the six cones generated by four of these
vertices, the corresponding four points lie on the same affine plane:

Exercise. Suppose for each of the eight vertices v there is a real
number r_v, and for each of the six large cones σ there is a vector
u_σ in $M_{\mathbb{R}} = \mathbb{R}^3$, such that $\langle u_\sigma, v \rangle = r_v$ whenever v is one of the
four vertices in σ. Show that there is then one vector u in $M_{\mathbb{R}}$
with $\langle u, v \rangle = r_v$ for all v. In particular, the points $p_v = (1/r_v) \cdot v$
cannot have each quadruple corresponding to a cone lying in a plane
unless all eight points are coplanar.[17]

A particularly important construction of toric varieties starts
with a rational polytope P in the dual space $M_{\mathbb{R}}$. We assume that P
is n-dimensional, but it is not necessary that it contain the origin.
From P a fan denoted Δ_P is constructed as follows. There is a cone
σ_Q of Δ_P for each face Q of P, defined by

$$\sigma_Q = \{ v \in N_{\mathbb{R}} : \langle u, v \rangle \leq \langle u', v \rangle \text{ for all } u \in Q \text{ and } u' \in P \} .$$

In other words, σ_Q is dual to the "angle" at Q consisting of all vectors
pointing from points of Q to points of P; this dual cone $\sigma_Q^\vee$ is
generated by vectors $u' - u$, where u and u' vary among vertices
of Q and P respectively. It is not hard to verify directly that these
cones form a fan. It is more instructive, however, to realize the fan as
a fan over a "dual polytope" in $N_{\mathbb{R}}$.

Proposition. *The cones* σ_Q, *as* Q *varies over the faces of* P,
form a fan Δ_P. *If* P *contains the origin as an interior point, then*
Δ_P *consists of cones over the faces of the polar polytope* P^o.

Proof. If the origin is an interior point, it is immediate from the
definition that σ_Q is the cone over the dual face Q^* of P^o, and the
second assertion follows. It follows from the definition that Δ_P is
unchanged when P is translated by some element u of M, or when
P is multiplied by a positive integer m: $\Delta_{mP+u} = \Delta_P$. Since any P
spanning $M_{\mathbb{R}}$ can be changed to one containing the origin as an
interior point by such translation and expansion, the first assertion
also follows.

Conversely, by the duality of polytopes, it follows that for a convex rational polytope K in $N_{\mathbb{R}}$ (containing the origin in its interior) the fan of cones over the faces of K is the same as Δ_P, where $P = K^\circ$ is its polar polytope. The toric variety $X(\Delta_P)$ will sometimes be denoted X_P. We look at some examples:

(1) If P is the simplex in $\mathbb{R}^n$ with vertices at the origin and the points $e_1, \ldots, e_n$, then Δ_P is the fan used to construct $\mathbb{P}^n$ as a toric variety: $X(\Delta_P) \cong \mathbb{P}^n$.

(2) If P is the cube in $\mathbb{R}^3$ with vertices at $\pm e_1^* \pm e_2^* \pm e_3^*$, then Δ_P is the fan over faces of the octahedron with vertices $\pm e_i$, and $X(\Delta_P) = \mathbb{P}^1 \times \mathbb{P}^1 \times \mathbb{P}^1$.

(3) Let P be the octahedron in $\mathbb{R}^3$ with vertices at the points $(\pm 1,0,0)$, $(0,\pm 1,0)$, and $(0,0,\pm 1)$, so the fan Δ_P is the fan over faces of the cube with vertices $(\pm 1,\pm 1,\pm 1)$. If N is taken to be the lattice of points $(x,y,z) \in \mathbb{Z}^3$ such that $x \equiv y \equiv z \pmod 2$, then any three of the four vertices of any face σ of the cube generate N, and the sums of opposite vertices of σ are equal. Each of the six corresponding open subvarieties U_σ has a singular point isomorphic to the cone over a quadric surface described in §1.3. We will come back to this example later.

CHAPTER 2

SINGULARITIES AND COMPACTNESS

2.1 Local properties of toric varieties

For any cone σ in a lattice N, the corresponding affine variety U_σ has a distinguished point, which we denote by x_σ. This point in U_σ is given by a map of semigroups

$$S_\sigma = \sigma^\vee \cap M \rightarrow \{1,0\} \subset \mathbb{C}^* \cup \{0\} = \mathbb{C},$$

defined by the rule

$$u \mapsto \begin{cases} 1 & \text{if } u \in \sigma^\perp \\ 0 & \text{otherwise} \end{cases}.$$

Note that this is well defined since $\sigma^\perp$ is a face of $\sigma^\vee$, which implies that the sum of two elements in $\sigma^\vee$ cannot be in $\sigma^\perp$ unless both are in $\sigma^\perp$.

Exercise. If σ spans $N_\mathbb{R}$, show that x_σ is the unique fixed point of the action of the torus T_N on U_σ. If σ does not span $N_\mathbb{R}$, show that there are no fixed points in U_σ.[1]

We first find the singular points of toric varieties. Suppose to start that σ spans $N_\mathbb{R}$, so $\sigma^\perp = \{0\}$. Let $A = A_\sigma$, $\mathfrak{m}$ the maximal ideal of A corresponding to the point x_σ, so $\mathfrak{m}$ is generated by all χ^u for nonzero u in S_σ. The square $\mathfrak{m}^2$ is generated by all χ^u for those u that are sums of two elements of $S_\sigma \smallsetminus \{0\}$. The *cotangent space* $\mathfrak{m}/\mathfrak{m}^2$ therefore has a basis of images of elements χ^u for those u in $S_\sigma \smallsetminus \{0\}$ that are not the sums of two such vectors. For example, the first elements in M lying along the edges of $\sigma^\vee$ are vectors of this kind. Suppose U_σ is nonsingular at the point x_σ. One characterization of nonsingularity is that the cotangent space $\mathfrak{m}/\mathfrak{m}^2$ is n-dimensional, since $\dim(U_\sigma) = \dim(T_N) = n$. This implies in particular that $\sigma^\vee$ cannot have more than n edges, and that the minimal generators along these edges must generate S_σ. Since S_σ generates M as a group, the minimal generators for S_σ must be a basis for M. The dual σ must therefore be generated by a basis for N. Hence U_σ is isomorphic to affine space $\mathbb{C}^n$.

A general σ has smaller dimension k. Let

$$N_\sigma = \sigma \cap N + (-\sigma \cap N)$$

be the sublattice of N generated (as a subgroup) by $\sigma \cap N$. Since σ is saturated, N_σ is also saturated, so the quotient group $N(\sigma) = N/N_\sigma$ is also a lattice. We may choose a splitting and write

$$N = N_\sigma \oplus N'' , \qquad \sigma = \sigma' \oplus \{0\} ,$$

where σ' is a cone in N_σ. Decomposing $M = M' \oplus M''$ dually, we have $S_\sigma = ((\sigma')^\vee \cap M') \oplus M''$, so

$$U_\sigma \cong U_{\sigma'} \times T_{N''} \cong U_{\sigma'} \times (\mathbb{C}^*)^{n-k} .$$

More intrinsically, the maps $N_\sigma \to N \to N(\sigma)$, $\sigma' \to \sigma \to \{0\}$, determine a fiber bundle

$$U_{\sigma'} \to U_\sigma \to T_{N(\sigma)}$$

that splits: $U_\sigma \cong U_{\sigma'} \times T_{N(\sigma)}$, but not canonically. If U_σ is nonsingular, $U_{\sigma'}$ must also be nonsingular, and the preceding discussion applies: σ' must be generated by a basis for N_σ. This proves:

Proposition. *An affine toric variety U_σ is nonsingular if and only if σ is generated by part of a basis for the lattice N, in which case*

$$U_\sigma \cong \mathbb{C}^k \times (\mathbb{C}^*)^{n-k} , \qquad k = \dim(\sigma) .$$

We therefore call a cone *nonsingular* if it is generated by part of a basis for the lattice, and we call a fan nonsingular if all of its cones are nonsingular, i.e., if the corresponding toric variety is nonsingular. Although a toric variety may be singular, every toric variety is *normal:*

Proposition. *Each ring $A_\sigma = \mathbb{C}[S_\sigma]$ is integrally closed.*

Proof. If σ is generated by $v_1, \ldots, v_r$, then $\sigma^\vee = \bigcap \tau_i^\vee$, where τ_i is the ray generated by $v_i \in \sigma$, so $A_\sigma = \bigcap A_{\tau_i}$. We have seen that each A_{τ_i} is isomorphic to $\mathbb{C}[X_1, X_2, X_2^{-1}, \ldots, X_n, X_n^{-1}]$, which is integrally closed, and the proposition follows from the fact that the intersection of integrally closed domains is integrally closed.

Exercise. (a) If S is any sub-semigroup of M, show that $\mathbb{C}[S]$ is integrally closed if and only if S is saturated.

(b) If S is a sub-semigroup of M, let $\tilde{S}$ be its saturation, i.e., $\tilde{S} = \{u \in M : p \cdot u \in S \text{ for some positive integer } p\}$. Show that $\mathbb{C}[\tilde{S}]$ is the integral closure of $\mathbb{C}[S]$, i.e., $\operatorname{Spec}(\mathbb{C}[\tilde{S}]) \to \operatorname{Spec}(\mathbb{C}[S])$ is the normalization map. Carry this out for S generated by $e_1^* + 2e_2^*$ and $2e_1^* + e_2^*$ in $M = \mathbb{Z}e_1^* + \mathbb{Z}e_2^*$.

(c) Deduce Gordon's lemma from the commutative algebra fact that the integral closure of a finitely generated domain over a field is finitely generated as a module over the domain.

Another important fact about toric varieties (but one we won't need here) is that they are all *Cohen-Macaulay* varieties: each of their local rings has depth n, i.e., contains a regular sequence of n elements, where n is the dimension of the local ring. We sketch the proof (assuming familiarity with properties of depth), referring to [Dani] for details. As above, it is enough to consider the case where $\dim(\sigma) = n$. In addition, we may replace the lattice N by any sublattice N' of finite index; this follows from the fact that, if σ' is the corresponding cone in N', the inclusion $A_\sigma \to A_{\sigma'}$ splits as a map of A_σ-modules, the splitting given by projecting $\sigma^\vee \cap M'$ onto $\sigma^\vee \cap M$.

If $\dim(\sigma) = n$, set $A = A_\sigma$, and $I = \oplus \mathbb{C} \chi^u$, the sum over all u in $\operatorname{Int}(\sigma^\vee) \cap M$. Orienting $\sigma^\vee$ and all its faces, one has an exact sequence

$$0 \to A/I \to C_{n-1} \to C_{n-2} \to \ldots \to C_1 \to C_0 \to 0 \, ,$$

where C_{n-k} is the direct sum of rings $\mathbb{C}[\sigma^\vee \cap \tau^\perp \cap M]$, for all k-dimensional faces τ of σ; the boundary maps are given by projections of faces in $\sigma^\vee$ onto smaller faces, with signs determined by the orientation. One knows by induction that C_{n-k} has depth $n-k$, and it follows that A/I has depth $n-1$. If I were a principal ideal, it would follow that A has depth n. If there is a u_0 in $\text{Int}(\sigma^\vee) \cap M$ such that $u - u_0$ is in $\sigma^\vee \cap M$ for all $u \in \text{Int}(\sigma^\vee) \cap M$, then $I = A \cdot \chi^{u_0}$. Although this need not be true, it can be achieved by replacing N by a sublattice N' of finite index.[2]

     The following exercise states a useful fact, although in situations where it comes up the conclusion can usually be seen by hand.

Exercise. If a torus T_N acts algebraically on a finite-dimensional vector space W, i.e., the map $T_N \to GL(W)$ is a morphism of affine algebraic groups, show that W decomposes into a direct sum of the spaces

$$W_u = \{w \in W : t \cdot w = \chi^u(t)w \text{ for all } t \in T_N\}$$

as u varies over M. [3]

Exercise. Deduce that for any algebraic action of a torus T on a finite dimensional vector space V, the ring of invariants $\text{Sym}(V)^T$ is Cohen-Macaulay.[4]

Exercise. Use the preceding exercise to show that the homogeneous coordinate ring of the Segre variety:

$$\mathbb{C}[X_1 Y_1, X_1 Y_2, \ldots, X_1 Y_q, X_2 Y_1, \ldots, X_2 Y_q, \ldots, X_p Y_1, \ldots, X_p Y_q]$$

is Cohen-Macaulay.

     More recently, Gubeladze has proved a conjecture of Anderson that the affine ring A_σ of a toric variety satisfies the generalization of a conjecture of Serre: all projective modules are free, or, in geometric language, all vector bundles on affine toric varieties are trivial.[5]

2.2   Surfaces; quotient singularities

Consider the case where $N = \mathbb{Z}^2$ and σ is generated by e_2 and $m e_1 - e_2$, generalizing the case $m = 2$ that we looked at earlier.

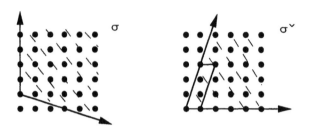

Then $A_\sigma = \mathbb{C}[S_\sigma] = \mathbb{C}[X, XY, XY^2, \ldots, XY^m]$. Setting $X = U^m$ and $Y = V/U$, we have

$$A_\sigma = \mathbb{C}[U^m, U^{m-1}V, \ldots, UV^{m-1}, V^m] \subset \mathbb{C}[U, V],$$

so $U_\sigma = \mathrm{Spec}(A_\sigma)$ is the *cone over the rational normal curve* of degree m. The inclusion of A_σ in $\mathbb{C}[U,V]$ corresponds to a mapping $\mathbb{C}^2 \to U_\sigma$.

The group $G = \mu_m = (m^{\mathrm{th}}$ roots of unity$) \cong \mathbb{Z}/m\mathbb{Z}$ acts on $\mathbb{C}^2$ by $\zeta\cdot(u,v) = (\zeta u, \zeta v)$, and U_σ is the quotient variety $\mathbb{C}^2/G$, i.e., U_σ is a *cyclic quotient singularity*. Algebraically, G acts on the coordinate ring $\mathbb{C}[U,V]$ by $F \mapsto F(\zeta U, \zeta V)$, and, via this action,

$$A_\sigma = \mathbb{C}[U,V]^G$$

is the ring of invariants.

The toric structure can be used to see this more naturally. Let $N' \subset N$ be the lattice generated by the generators e_2 and $me_1 - e_2$ of σ, and let σ' be the same cone as σ, but regarded in N' (since $N'_\mathbb{R} = N_\mathbb{R}$). Since σ' is generated by two generators for N', $U_{\sigma'} = \mathbb{C}^2$, and the inclusion of N' in N gives a map $\mathbb{C}^2 = U_{\sigma'} \to U_\sigma$. The claim is that this is the same as the map constructed by hand above. To see this, note that N' is generated by me_1 and e_2, so $M' \supset M$ is generated by $\frac{1}{m}e_1^*$ and e_2^*, corresponding to monomials U and Y, with $U^m = X$. The generators for $S_{\sigma'}$ are $\frac{1}{m}e_1^*$ and $\frac{1}{m}e_1^* + e_2^*$, so $A_{\sigma'} = \mathbb{C}[U, UY] = \mathbb{C}[U,V]$, with $V = UY$, and we recover the previous description.

A similar procedure applies to an arbitrary singular two-dimensional affine toric variety. First note that, with an appropriate choice of basis for N, we may assume the cone σ has the form

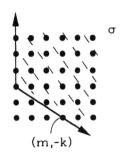 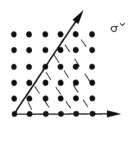

for some $0 < k < m$, with k and m relatively prime integers. To prove this, note that any minimal generator along an edge of σ is part of a basis for $N = \mathbb{Z}^2$, so we can take one to be $(0,1)$, and the other (m,x), for m a positive integer. Applying an automorphism of the lattice,

$$\begin{pmatrix} 1 & 0 \\ c & 1 \end{pmatrix} \cdot \begin{pmatrix} m & 0 \\ x & 1 \end{pmatrix} = \begin{pmatrix} m & 0 \\ cm+x & 1 \end{pmatrix},$$

we see that x can be changed arbitrarily modulo m, so we can take $x = -k$, $0 \le k < m$. Of course, if $x \equiv 0 \pmod m$, then σ is generated by a basis for N, and we are in the nonsingular case. That k and m are relatively prime comes from the fact that $(m,-k)$ is taken to be a minimal generator along the edge.

With σ as above, $A_\sigma = \oplus \mathbb{C} \cdot X^i Y^j$, the sum over (i,j) with $j \le \frac{m}{k} i$. Let N' be generated by $m e_1 - k e_2$ and e_2, i.e. by $m e_1$ and e_2. Then as before, M' is generated by $\frac{1}{m} e_1^*$ and e_2^*, corresponding to monomials U and Y, and the corresponding cone σ' has $S_{\sigma'}$ generated by $\frac{1}{m} e_1^*$ and $k \cdot \frac{1}{m} e_1^* + e_2^*$. Therefore $A_{\sigma'}$ is $\mathbb{C}[U, U^k Y] = \mathbb{C}[U,V]$, with $V = U^k Y$, so $U_{\sigma'} = \mathbb{C}^2$.

The group $G = \mu_m$ acts on $U_{\sigma'} = \mathbb{C}^2$ by $\zeta \cdot (u,v) = (\zeta u, \zeta^k v)$, and

$$U_\sigma = U_{\sigma'}/G = \mathbb{C}^2/G.$$

Equivalently, $A_\sigma = (A_{\sigma'})^G$. In fact, G acts on the larger ring $\mathbb{C}[M'] = \mathbb{C}[U, U^{-1}, V, V^{-1}] = \mathbb{C}[U, U^{-1}, Y, Y^{-1}]$, by $U \mapsto \zeta U$, $Y \mapsto Y$, and the ring of invariants is $\mathbb{C}[X, X^{-1}, Y, Y^{-1}] = \mathbb{C}[M]$. Therefore

$$A_\sigma = A_{\sigma'} \cap \mathbb{C}[M] = A_{\sigma'} \cap (\mathbb{C}[M'])^G = (A_{\sigma'})^G.$$

We will describe A_σ more explicitly in §2.6.

More generally still, and more intrinsically, for a lattice N of any rank, if $N' \subset N$ is a sublattice of finite index, let $M \subset M'$ be the dual lattices. There is the canonical duality pairing

$$M'/M \times N/N' \longrightarrow \mathbb{Q}/\mathbb{Z} \hookrightarrow \mathbb{C}^* ,$$

the first map by the pairing $\langle\,,\,\rangle$, the second by $q \mapsto \exp(2\pi i q)$. Now $G = N/N'$ acts on $\mathbb{C}[M']$ by

$$v \cdot \chi^{u'} = \exp(2\pi i \langle u', v \rangle) \cdot \chi^{u'}$$

for $v \in N$, $u' \in M'$. And, via this natural action,

$$(*) \qquad\qquad \mathbb{C}[M']^G = \mathbb{C}[M] .$$

Hence $G = N/N'$ acts on the torus $T_{N'}$ and $T_{N'}/G = T_N$. To prove $(*)$, it suffices to take a basis $e_1, \ldots, e_n$ for N so that $m_1 e_1, \ldots, m_n e_n$ are generators for N', for some positive integers m_i. Then $\mathbb{C}[M']$ is the Laurent polynomial ring in generators X_i, and $\mathbb{C}[M]$ is the Laurent polynomial ring in generators U_i, with $(U_i)^{m_i} = X_i$. An element $(a_1, \ldots, a_n)$ in $\oplus \mathbb{Z}/m_i\mathbb{Z} = N/N'$ acts on monomials by multiplying $U_1^{\ell_1} \cdot \ldots \cdot U_n^{\ell_n}$ by $\exp(2\pi i(\Sigma a_i \ell_i/m_i))$, from which $(*)$ follows at once. In the special case when N has rank 2, with basis e_1 and e_2, and N' is generated by $m e_1$ and e_2, N/N' is isomorphic to μ_m, with the image of e_1 corresponding to $\zeta = \exp(2\pi i/m)$, and one checks that the general action of N/N' specializes to the above action of μ_m.

Now suppose σ is an n-simplex in N, i.e., σ is generated by n independent vectors. Let $N' \subset N$ be the sublattice generated by the minimal elements in $\sigma \cap N$ along its n edges. As before, this gives a cone σ' in N', with $\mathbb{C}^n = U_{\sigma'} \to U_\sigma$. The abelian group $G = N/N'$ acts on $U_{\sigma'}$, with

$$U_\sigma = U_{\sigma'}/G = \mathbb{C}^n/G .$$

This follows as before from the case of the torus, by intersecting the ring $A_{\sigma'}$ with $\mathbb{C}[M']^G = \mathbb{C}[M]$.

If σ is any simplex, i.e., generated by independent vectors, then U_σ is a product of a quotient as above and a torus. In particular, if Δ is a *simplicial fan*, — all the cones in Δ are simplices — then $X(\Delta)$ is an *orbifold* or *V-manifold*, i.e., it has only quotient singularities.[6]

Exercise. Let σ be the cone generated by

$$e_1, \; e_2, \; \cdots \; , \; e_{n-1}, \; -e_1 - e_2 - \cdots - e_{n-1} + m e_n \; ,$$

where $e_1, \ldots, e_n$ is a basis for N. Show that

(i) $U_\sigma = \mathbb{C}^n/\mu_m$, where the mth roots of unity μ_m act by
$\zeta \cdot (x_1, \ldots, x_n) = (\zeta \cdot x_1, \ldots, \zeta \cdot x_n)$;

(ii) U_σ is the cone over the m-tuple Veronese embedding of $\mathbb{P}^{n-1}$.

Exercise. Show that if m and $a_1, \ldots, a_n$ are positive integers, the quotient of $\mathbb{C}^n$ by the cyclic group μ_m acting by

$$\zeta \cdot (z_1, \ldots, z_n) \;=\; (\zeta^{a_1} z_1, \ldots, \zeta^{a_n} z_n)$$

can be constructed as an affine toric variety U_σ, by taking

$$N' \;=\; \sum_{i=1}^{n} \mathbb{Z} \cdot (1/a_i) \cdot e_i \;\subset\; N \;=\; N' + \mathbb{Z} \cdot (1/m) \cdot (e_1 + \ldots + e_n) \; ,$$

and the cone σ generated by $e_1, \ldots, e_n$. When $a_1 = \ldots = a_n = 1$, show that this agrees with the construction of the preceding exercise.

One can sometimes extend this construction to non-affine toric varieties by making the group actions compatible on affine open subvarieties. An interesting example is the *twisted* or *weighted projective space* $\mathbb{P}(d_0, \ldots, d_n)$, where $d_0, \ldots, d_n$ are any positive integers. To construct this as a toric variety, start with the same fan used in the construction of projective space, i.e., its cones generated by proper subsets of $\{v_0, v_1, \ldots, v_n\}$, where any n of these vectors are linearly independent, and their sum is zero; however, the lattice N is taken to be generated by the vectors $(1/d_i) \cdot v_i$, $0 \le i \le n$. The resulting toric variety is in fact the variety

$$\mathbb{P}(d_0, \ldots, d_n) \;=\; \mathbb{C}^{n+1} \smallsetminus \{0\} \,/\, \mathbb{C}^* \; ,$$

where $\mathbb{C}^*$ acts by $\zeta \cdot (x_0, \ldots, x_n) = (\zeta^{d_0} x_0, \ldots, \zeta^{d_n} x_n)$.

Exercise. (a) With σ_i the cone generated by the complement of e_i, use the preceding exercise to identify U_{σ_i} with a quotient of $\mathbb{C}^n$, and use the standard map $(x_1, \ldots, x_n) \mapsto (x_1 : \ldots : x_{i-1} : 1 : x_i : \ldots : x_n)$ to identify this quotient with the open set U_i in $\mathbb{P}(d_0, \ldots, d_n)$ consisting of points whose i^{th} coordinate is nonzero.

(b) Write this twisted projective space as a quotient of a nonsingular toric variety by a finite abelian group.[7]

For any cone in a simplicial fan, one can find a sublattice to write the corresponding open subvariety of the toric variety as a quotient by a finite abelian group. Sometimes, as in the example of twisted projective spaces, one can find one sublattice that works for all these open sets at once, but this is not always possible:

Exercise. Let Δ be the fan in $N = \mathbb{Z}^2$, covering $\mathbb{R}^2$, with edges generated by the vectors $-e_1$, $-e_1 - e_2$, $-e_2$, and $2e_1 + 3e_2$. Show that for every sublattice N' of finite index in N, with Δ' the corresponding fan in N', the variety $X(\Delta')$ is singular.

2.3 One-parameter subgroups; limit points

In this section, we use one-parameter subgroups of the torus, and their limit points in toric varieties, to see how to recover the fan from the torus action.

For each integer k we have a homomorphism of algebraic groups

$$\mathbb{G}_m \rightarrow \mathbb{G}_m , \quad z \mapsto z^k .$$

Here we write $\mathbb{G}_m$ for the multiplicative algebraic group, i.e., for $\mathbb{C}^*$.

Exercise. Show that these are all of them:

$$\mathrm{Hom}_{\text{alg. gp.}}(\mathbb{G}_m, \mathbb{G}_m) = \mathbb{Z} . \quad [8]$$

Given a lattice N, with dual M, we have the corresponding torus

$$T_N = \mathrm{Hom}(M, \mathbb{G}_m) .$$

From the preceding exercise, it follows (by taking a basis for N) that

$$\mathrm{Hom}(\mathbb{G}_m, T_N) = \mathrm{Hom}(\mathbb{Z}, N) = N .$$

This means that every *one-parameter subgroup* $\lambda: \mathbb{G}_m \rightarrow T_N$ is given by a unique v in N. Let λ_v denote the one-parameter subgroup corresponding to v. For $z \in \mathbb{C}^*$, $\lambda_v(z)$ is in T_N, so is given by a group homomorphism from M to $\mathbb{C}^*$. Explicitly, for u in M,

$$\lambda_v(z)(u) = \chi^u(\lambda_v(z)) = z^{\langle u, v \rangle} ,$$

where $\langle \, , \rangle$ is the dual pairing $M \otimes N \rightarrow \mathbb{Z}$.

Dually,

$$\text{Hom}(T_N, \mathbb{G}_m) = \text{Hom}(N, \mathbb{Z}) = M .$$

Every *character* $\chi: T_N \rightarrow \mathbb{G}_m$ is given by a unique u in M. The character corresponding to u can be identified with the function χ^u in the coordinate ring $\mathbb{C}[M] = \Gamma(T_N, \mathbb{O}^*)$.

Exercise. (a) Show that for lattices N and N', the mapping

$$\text{Hom}_{\mathbb{Z}}(N', N) \rightarrow \text{Hom}_{\text{alg. gp.}}(T_{N'}, T_N) , \quad \varphi \mapsto \varphi_* ,$$

is an isomorphism. (A mapping between tori induces a corresponding mapping on one-parameter subgroups.)

(b) Composition gives a pairing

$$\text{Hom}(T_N, \mathbb{G}_m) \times \text{Hom}(\mathbb{G}_m, T_N) \rightarrow \text{Hom}(\mathbb{G}_m, \mathbb{G}_m) .$$

Show that, by the above identifications, this is the duality pairing

$$\langle \, , \rangle : M \times N \rightarrow \mathbb{Z} .$$

Note in particular that the above prescription shows how to recover the lattice N from the torus T_N. Given a cone σ, we next want to see how to recover σ from the torus embedding $T_N \subset U_\sigma$.

The key is to look at *limits* $\lim_{z \to 0} \lambda_v(z)$ for various $v \in N$, as the complex variable z approaches the origin.[9] For example, suppose σ is generated by part of a basis $e_1, \ldots, e_k$ for N, so U_σ is $\mathbb{C}^k \times (\mathbb{C}^*)^{n-k}$. For $v = (m_1, \ldots, m_n) \in \mathbb{Z}^n$, $\lambda_v(z) = (z^{m_1}, \ldots, z^{m_n})$. Then $\lambda_v(z)$ has a limit in U_σ if and only if all m_i are nonnegative and $m_i = 0$ for $i > k$. In other words, the limit exists exactly when v is in σ. In this case, the limit is $(\delta_1, \ldots, \delta_n)$, where $\delta_i = 1$ if $m_i = 0$ and $\delta_i = 0$ if $m_i > 0$. Each of these limit points is one of the distinguished points x_τ for some face τ of σ.

In general, for each cone τ in a fan Δ, we have defined the distinguished point x_τ in U_τ. If τ is a face of σ, then U_τ is contained in U_σ, so we must be able to realize x_τ as a homomorphism of semigroups from S_σ to $\mathbb{C}$. This homomorphism is

$$S_\sigma \rightarrow \{1,0\} \subset \mathbb{C}, \quad u \mapsto \begin{cases} 1 & \text{if } u \in \tau^\perp \\ 0 & \text{otherwise} \end{cases},$$

which is well defined since $\tau^\perp \cap \sigma^\vee$ is a face of $\sigma^\vee$. From this it follows that the resulting point x_τ of $X(\Delta)$ is independent of σ; that is, if $\tau < \sigma < \gamma$, then the inclusion of U_σ in U_γ takes the point defined in U_σ to the point defined in U_γ. We note also that these points are all distinct; this follows from the fact that x_τ is not in U_σ if σ is a proper face of τ. As we will see later, there is exactly one such point in each orbit of T_N on $X(\Delta)$.

Claim 1. *If v is in $|\Delta|$, and τ is the cone of Δ that contains v in its relative interior, then $\lim_{z \to 0} \lambda_v(z) = x_\tau$.*

For the proof, look in U_σ for any σ containing τ as a face, and identify $\lambda_v(z)$ with the homomorphism from M to $\mathbb{C}^*$ that takes u to $z^{\langle u,v \rangle}$. For u in S_σ, we have $\langle u,v \rangle \geq 0$, with equality exactly when u belongs to $\tau^\perp$. It follows that the limiting homomorphism from $S_\sigma \subset M$ to $\mathbb{C}$ is precisely that which defines x_τ. (One should check that this is the topological limit, say by choosing m generators χ^u for S_σ to embed U_σ in $\mathbb{C}^m$.)

Claim 2. *If v is not in any cone of Δ, then $\lim_{z \to 0} \lambda_v(z)$ does not exist in $X(\Delta)$.*

In fact, if v is not in σ, the points $\lambda_v(z)$ have no limit points in U_σ as z approaches 0. To see this, take u in $\sigma^\vee$ with $\langle u,v \rangle < 0$ (possible since $\sigma = (\sigma^\vee)^\vee$). Then $\chi^u(\lambda_v(z)) = z^{\langle u,v \rangle} \to \infty$ as $z \to 0$.

With these claims, $\sigma \cap N$ is characterized as the set of v for which $\lambda_v(z)$ has a limit in U_σ as $z \to 0$, and the limit is x_σ if v is in the relative interior of σ. For those v not in the *support* $|\Delta|$ of Δ, i.e., the union of the cones in Δ, there is no limit (or converging subsequence).

Exercise. For $v \in N$, show that $\lambda_v : \mathbb{C}^* \to T_N$ extends to a morphism from $\mathbb{C}$ to $X(\Delta)$ if and only if v belongs to $|\Delta|$, and λ_v extends to a morphism from $\mathbb{P}^1$ to $X(\Delta)$ if and only if v and $-v$ belong to $|\Delta|$.

2.4 Compactness and properness

Recall that a complex variety is compact in its classical topology exactly when it is complete (proper) as an algebraic variety. For a toric variety, we can see this in terms of the fan:

A toric variety $X(\Delta)$ is compact if and only if its support $|\Delta|$ is the whole space $N_{\mathbb{R}}$.

Because of this we say that a fan Δ is *complete* if $|\Delta| = N_{\mathbb{R}}$. One implication is easy: if the support were not all of $N_{\mathbb{R}}$, since Δ is finite, there would be a lattice point v not in any cone; the fact that $\lambda_v(z)$ has no limit point as $z \to 0$ contradicts compactness.

Before proving the converse, we state the appropriate generalization. Let $\varphi \colon N' \to N$ be a homomorphism of lattices that maps a fan Δ' into a fan Δ as in §1.4, so defining a morphism $\varphi_* \colon X(\Delta') \to X(\Delta)$.

Proposition. *The map $\varphi_* \colon X(\Delta') \to X(\Delta)$ is proper if and only if $\varphi^{-1}(|\Delta|) = |\Delta'|$.*

A variety is compact when the map to a point is proper, so the preceding case is recovered by taking the second lattice to be $\{0\}$.

Proof of the proposition. $\Rightarrow$: If v' is in N' but in no cone of Δ', and $v = \varphi(v')$ is in a cone of Δ, then $\varphi_*(\lambda_{v'}(z)) = \lambda_v(z)$ has a limit in $X(\Delta)$, but $\lambda_{v'}(z)$ has no converging subsequences as $z \to 0$, which contradicts properness.

$\Leftarrow$: We use the *valuative criterion of properness:* a morphism $f \colon X \to Y$ of varieties (or a separated morphism of schemes of finite type) is proper if and only if for any discrete valuation ring R, with quotient field K, any commutative diagram

$$\bigodot = \text{Spec}(K) \longrightarrow X$$

$$\bigodot = \text{Spec}(R) \longrightarrow Y$$

can be filled in (uniquely) as shown. When X is irreducible, one may assume the image of the map $\text{Spec}(K) \to X$ is in a given open subset U of X.[10]

Apply this with $X = X(\Delta') \supset U = T_{N'}$, $Y = X(\Delta)$, $f = \varphi_*$; assume $\text{Spec}(R)$ maps to U_σ. The map from $\text{Spec}(K)$ to U is given by a homomorphism $\alpha: M' \to K^*$ of groups. We want to find σ' mapping to σ so we can fill in the diagram

$$K \longleftarrow \mathbb{C}[M'] \supset \mathbb{C}[S_{\sigma'}]$$

$$R \longleftarrow \mathbb{C}[S_\sigma]$$

The fact that $\text{Spec}(R)$ maps to U_σ says that, if ord is the order function of the discrete valuation, then $\text{ord} \circ \alpha \circ \varphi^*$ is nonnegative on $\sigma^\vee \cap M$; equivalently,

$$\varphi(\text{ord} \circ \alpha) = \text{ord} \circ \alpha \circ \varphi^* \in (\sigma^\vee)^\vee = \sigma .$$

By the assumption, there is a cone σ' so that $\varphi(\sigma') \subset \sigma$ and $\text{ord} \circ \alpha \in \sigma'$. This says that $\text{ord} \circ \alpha$ is nonnegative on $\sigma'^\vee$, which is precisely the condition needed to fill in the diagram.

A fundamental example of a proper map is *blowing up*. We saw the first example of this in the construction of the blow-up of $U_\sigma = \mathbb{C}^2$ at the origin $x_\sigma = (0,0)$ in Chapter 1. More generally, suppose a cone σ in Δ is generated by a basis $v_1, \ldots, v_n$ for N. Set $v_0 = v_1 + \ldots + v_n$, and replace σ by the cones that are generated by those subsets of $\{v_0, v_1, \ldots, v_n\}$ not containing $\{v_1, \ldots, v_n\}$. This gives a fan Δ', and the resulting proper map $X(\Delta') \to X(\Delta)$ is the blow-up of $X(\Delta)$ at the point x_σ. To see this, since nothing is changed except over U_σ, we may assume Δ consists of σ and all of its faces, so $X(\Delta) = U_\sigma = \mathbb{C}^n$, with x_σ the origin, and $v_i = e_i$ for $1 \leq i \leq n$. Then $X(\Delta')$ is covered by the open affine varieties U_{σ_i}, where σ_i is the cone generated by $e_0, e_1, \ldots, \hat{e}_i, \ldots, e_n$, $1 \leq i \leq n$, and $\sigma_i^\vee$ is generated by $e_i^*, e_1^* - e_i^*, \ldots, e_n^* - e_i^*$. The corresponding coordinate ring is

$$A_{\sigma_i} = \mathbb{C}[X_i, X_1 X_i^{-1}, \ldots, X_n X_i^{-1}].$$

On the other hand, the blow-up of the origin in $\mathbb{C}^n$ is the subvariety of $\mathbb{C}^n \times \mathbb{P}^{n-1}$ defined by the equations $X_i T_j = X_j T_i$, where $T_1, \ldots, T_n$ are homogeneous coordinates on $\mathbb{P}^{n-1}$. The set U_i where $T_i \neq 0$ has $X_j = X_i \cdot (T_j / T_i)$, so U_i is $\mathbb{C}^n$, with coordinates X_i and $T_j / T_i = X_j / X_i$ for $j \neq i$. So $U_i = U_{\sigma_i}$ with σ_i as above, and with the same gluing.

Exercise. For a nonsingular affine toric variety of the form $\mathbb{C}^k \times (\mathbb{C}^*)^{n-k}$, show how to construct the blow-up along $(0) \times (\mathbb{C}^*)^{n-k}$ as a toric variety.

Exercise. If Δ is an infinite fan, show that $X(\Delta)$ is never compact, even if $|\Delta| = N_\mathbb{R}$ and all $\lambda_v(z)$ have limits in $X(\Delta)$. Why does the criterion for properness not apply? Give an example of an infinite fan Δ in $\mathbb{Z}^2$ whose support is all of $\mathbb{R}^2$. Show that, for infinite fans, the proposition remains true with the added condition that there are only finitely many cones in Δ' mapping to a given cone in Δ. [11]

Exercise. *(Fiber bundles)* Suppose $0 \to N' \to N \to N'' \to 0$ is an exact sequence of lattices, and suppose Δ', Δ, and Δ'' are fans in N', N, and N'' that are compatible with these mappings as in §1.4, giving rise to maps

$$X(\Delta') \to X(\Delta) \to X(\Delta'').$$

Suppose there is a fan $\widetilde{\Delta}''$ in N that lifts Δ'', i.e., each cone in Δ'' is the isomorphic image of a unique cone in $\widetilde{\Delta}''$, such that the cones σ in Δ are of the form

$$\sigma = \sigma' + \sigma'' = \{v' + v'' : v' \in \sigma', v'' \in \sigma''\}$$

for σ' a cone in Δ' and σ'' a cone in $\widetilde{\Delta}''$. Show that the above sequence is a locally trivial fibration.

Exercise. Construct the projective bundle

$$\mathbb{P}(\mathcal{O}(a_1) \oplus \mathcal{O}(a_2) \oplus \ldots \oplus \mathcal{O}(a_r)) \;\to\; \mathbb{P}^n$$

as a toric variety.[12]

Exercise. Let $\varphi_*: X(\Delta') \to X(\Delta)$ be the map arising from a homomorphism of lattices $\varphi: N' \to N$ mapping Δ' to Δ as in §1.4. Show that for a cone σ' in Δ', φ_* maps the point $x_{\sigma'}$ of $X(\Delta')$ to the point x_σ of $X(\Delta)$, where σ is the smallest cone of Δ that contains σ'.[13]

Exercise. Find a subdivision of the cone generated by $(1,0,0)$, $(0,1,0)$, and $(0,0,1)$ in $\mathbb{Z}^3$, including the ray through $(1,1,1)$, such that: (i) the resulting toric variety X' is nonsingular; (ii) the resulting proper map $X' \to \mathbb{C}^3$ is an isomorphism over $\mathbb{C}^3 \smallsetminus \{0\}$; but (iii) this map does not factor through the blow-up of $\mathbb{C}^3$ at the origin. In particular, this map does not factor into a composite of blow-ups along smooth centers.[14]

2.5 Nonsingular surfaces

Let us see what *two-dimensional nonsingular complete toric varieties* look like. They are given by specifying a sequence of lattice points

$$v_0, \; v_1, \; \ldots \; , \; v_{d-1}, \; v_d = v_0$$

in counterclockwise order, in $N = \mathbb{Z}^2$, such that successive pairs generate the lattice.

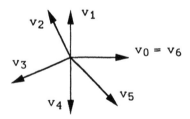

From the fact that v_0 and v_1 are a basis for the lattice, and v_1 and v_2 are also a basis, we know that $v_2 = -v_0 + a_1 v_1$ for some integer a_1. In general, we must have

$$a_i v_i = v_{i-1} + v_{i+1} , \quad 1 \le i \le d ,$$

for some integers a_i. We will discuss later the conditions these integers must satisfy.

The possible configurations are topologically constrained. For example, two of the cones cannot be arranged with v_j in the angle strictly between v_{i+1} and $-v_i$ and v_{j+1} in the angle strictly between $-v_i$ and $-v_{i+1}$:

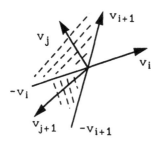

Exercise. Prove this.[15]

We want to classify all of these surfaces. The cases when the number d of edges is small are easy to do by hand:

Exercise. Show that for $d = 3$, the resulting toric variety must be $\mathbb{P}^2$, and for $d = 4$, one gets a Hirzebruch surface $\mathbb{F}_a$, both as constructed in §1.1.

Given one of these toric surfaces, we know how to construct another that is the blow-up of the first at a T_N-fixed point: simply insert the sum of two adjacent vectors. In fact:

Proposition. *All complete nonsingular toric surfaces are obtained from* $\mathbb{P}^2$ *or* $\mathbb{F}_a$ *by a succession of blow-ups at* T_N-*fixed points.*

This follows from the preceding exercise and the following:

Claim. *If* $d \ge 5$, *there must be some* j, $1 \le j \le d$, *such that* v_{j-1} *and* v_{j+1} *generate a strongly convex cone, and*

$$v_j = v_{j-1} + v_{j+1} .$$

The proof of this claim is outlined in the following exercise.

Exercise. (a) Show that if $d \geq 4$ there must be two opposite vectors in the sequence, i.e., $v_j = -v_i$ for some i, j.
   (b) Suppose $v_i = -v_0$ and $i \geq 3$. Show that $v_j = v_{j-1} + v_{j+1}$ for some $0 < j < i$. (16)

Exercise. Show that the integers $a_1, \ldots, a_d$ must satisfy the equation

(*) $$\begin{pmatrix} 0 & -1 \\ 1 & a_1 \end{pmatrix} \cdot \begin{pmatrix} 0 & -1 \\ 1 & a_2 \end{pmatrix} \cdots \begin{pmatrix} 0 & -1 \\ 1 & a_d \end{pmatrix} = \begin{pmatrix} 1 & 0 \\ 0 & 1 \end{pmatrix}. \quad (17)$$

Exercise. Show that inserting $v' = v_k + v_{k+1}$ between v_k and v_{k+1} changes the sequence of integers $a_1, \ldots, a_d$ by adding 1 to a_k and a_{k+1} and inserting a 1 between them. Deduce from this and the preceding discussion that the integers $a_1, \ldots, a_d$ must satisfy the equation

(**) $a_1 + a_2 + \ldots + a_d = 3d - 12$.

Exercise. Show conversely that any integers $a_1, \ldots, a_d$ satisfying (*) and (**) arise from a nonsingular two-dimensional toric variety. Show that the sequence $0, 2, 1, 3, 1, 3, 1, 1$ satisfies these conditions, and describe the corresponding surface. (18)

Exercise. Each v_i determines a curve $D_i \cong \mathbb{P}^1$ in X (as in §1.1). Show that the normal bundle to this embedding is the line bundle $\mathcal{O}(-a_i)$ on $\mathbb{P}^1$. Show that successive curves meet transversally, but are otherwise disjoint: (19)

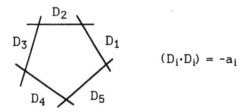

$$(D_i \cdot D_i) = -a_i$$

The classification of smooth projective toric varieties of higher dimension is an active problem. (20)

2.6   Resolution of singularities

In §2.2 we fixed the cones and refined the lattice. Now we fix the lattice and subdivide the cones. Suppose Δ' is a *refinement* of Δ, i.e., each cone of Δ is a union of cones in Δ'. The morphism $X(\Delta') \to X(\Delta)$ induced by the identity map of N is *birational* and *proper;* indeed, it is an isomorphism on the open torus T_N contained in each, and it is proper by the proposition in §2.4.

This construction can be used on singular toric varieties to resolve singularities. Consider the example where σ is the cone in $\mathbb{Z}^2$ generated by $3e_1 - 2e_2$ and e_2, and insert the edges through the points e_1 and $2e_1 - e_2$. The indicated subdivision

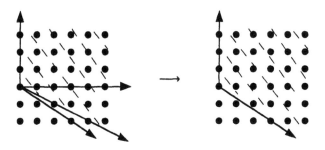

gives a nonsingular $X(\Delta')$ mapping birationally and properly to U_σ.

This can be generalized to any two-dimensional toric singularity. Given a cone σ that is not generated by a basis for N, we have seen that we can choose generators e_1 and e_2 for N so that σ is generated by $v = e_2$ and $v' = me_1 - ke_2$, $0 < k < m$, with k and m relatively prime. Insert the line through e_1:

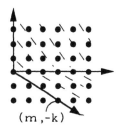

$(m, -k)$

The cone generated by e_1 and e_2 corresponds to a nonsingular open set, while the other cone generated by e_1 and $me_1 - ke_2$ corresponds

to a variety whose singular point is less singular than the original one.
To see this, rotate the picture by 90°, moving e_1 to e_2, and then
translate the other basic vector vertically (by a matrix $\begin{pmatrix} 1 & 0 \\ c & 1 \end{pmatrix}$ as in
§2.2) to put it in the position $(m_1, -k_1)$, with $m_1 = k$, $0 \le k_1 < m_1$,
and $k_1 = a_1 k - m$ for some integer $a_1 \ge 2$:

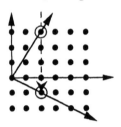

This corresponds to a smooth cone when $k_1 = 0$. Otherwise
$m/k = a_1 - k_1/m_1 = a_1 - 1/(m_1/k_1)$, and the process can be repeated.
The process continues as in the Euclidean algorithm, or in the
construction of the continued fraction, but with alternating signs:

$$\frac{m}{k} = a_1 - \cfrac{1}{a_2 - \cfrac{1}{\cdots - \cfrac{1}{a_r}}}$$

with integers $a_i \ge 2$. This is called the *Hirzebruch-Jung continued
fraction* of m/k.

Exercise. (a) Show that the edges drawn in the above process are
exactly those through the vertices on the edge of the polygon that is
the convex hull of the nonzero points in $\sigma \cap N$:

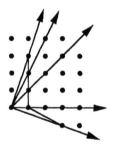

(b) Show that there are r added vertices $v_1, \ldots, v_r$ between the given vertices $v = v_0$ and $v' = v_{r+1}$, and $a_i v_i = v_{i-1} + v_{i+1}$.

(c) Show that these added rays correspond to exceptional divisors $E_i \cong \mathbb{P}^1$, forming a chain

with self-intersection numbers $(E_i \cdot E_i) = -a_i$.

(d) Show that $\{v_0, \ldots, v_{r+1}\}$ is a minimal set of generators of the semigroup $\sigma \cap N$.

Exercise. Show that the algebra $A_\sigma = \mathbb{C}[S_\sigma]$ has a minimal set of generators $\{u^{s_i} v^{t_i}, \ 1 \le i \le e\}$, where the embedding dimension e and the exponents are determined as follows. Let $b_2, \ldots, b_{e-1}$ be the integers (each at least 2) arising in the Hirzebruch-Jung continued fraction of $m/(m-k)$. Then

$$s_1 = m, \quad s_2 = m - k, \quad s_{i+1} = b_i \cdot s_i - s_{i-1} \quad \text{for } 2 \le i \le e-1;$$
$$t_1 = 0, \quad t_2 = 1, \quad t_{i+1} = b_i \cdot t_i - t_{i-1} \quad \text{for } 2 \le i \le e-1. \tag{21}$$

Exercise. Let σ be the cone generated by e_2 and $(k+1)e_1 - ke_2$. Show that S_σ is generated by $u_1 = e_1^*$, $u_2 = k e_1^* + (k+1)e_2^*$, and $u_3 = e_1^* + e_2^*$, with $(k+1)u_3 = u_1 + u_2$. Deduce that

$$A_\sigma = \mathbb{C}[Y_1, Y_2, Y_3]/(Y_3^{k+1} - Y_1 Y_2),$$

which is the *rational double point of type* A_k. Show that the resolution of singularities given by the above toric construction has k exceptional divisors in a chain, each isomorphic to $\mathbb{P}^1$ and with self-intersection -2. [22]

Exercise. Let σ be generated by e_2 and $me_1 - ke_2$ as above, and let σ' be generated by e_2 and $m'e_1 - k'e_2$, with $0 < k' < m'$ relatively prime. Show that $U_{\sigma'}$ is isomorphic to U_σ if and only if $m' = m$, and $k' = k$ or $k' \cdot k \equiv 1 \pmod{m}$. [23]

Given a fan Δ in any lattice N, and any lattice point v in N, one can subdivide Δ to a fan Δ' as follows: each cone that contains v is replaced by the joins (sums) of its faces with the ray through v; each cone not containing v is left unchanged.

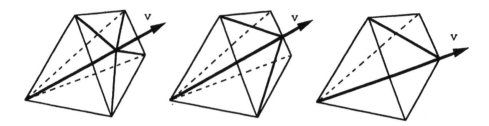

Since Δ' has the same support as Δ, the induced mapping from $X(\Delta')$ to $X(\Delta)$ is proper and birational. The goal is to choose a succession of such subdivisions to get to a nonsingular toric variety.

Exercise. Show that one can subdivide any fan, by successively adding vectors in larger and larger cones, until it becomes simplicial.

Now if σ is a k-dimensional simplicial cone, and $v_1, \ldots, v_k$ are the first lattice points along the edges of σ, the *multiplicity* of σ is defined to be the index of the lattice generated by the v_i in the lattice generated by σ:

$$\text{mult}(\sigma) \; = \; [N_\sigma : \mathbb{Z}v_1 + \ldots + \mathbb{Z}v_k] \,.$$

Note that U_σ is nonsingular precisely when the multiplicity of σ is one.

Exercise. Show that if $\text{mult}(\sigma) > 1$ there is a lattice point of the form $v = \Sigma t_i v_i$, $0 \leq t_i < 1$. For such v, taken minimal along its ray, show that the multiplicities of the subdivided k-dimensional cones are $t_i \cdot \text{mult}(\sigma)$, with one such cone for each nonzero t_i. [24]

From the preceding two exercises, one has a procedure for resolving the singularities of any toric variety — never leaving the world of toric varieties:

Proposition. *For any toric variety $X(\Delta)$, there is a refinement $\widetilde{\Delta}$ of Δ so that $X(\widetilde{\Delta}) \to X(\Delta)$ is a resolution of singularities.*

In particular, the resulting resolution is *equivariant*, i.e., the map commutes with the action of the torus.

Exercise. For surfaces, show that this procedure is the same as that described at the beginning of this section. Show that the integer a_i found there is the multiplicity of the cone generated by v_{i-1} and v_{i+1}.

Let N be a lattice of rank 3, with σ the cone generated by vectors v_1, v_2, v_3, and v_4 that generate N as a lattice and satisfy $v_1 + v_3 = v_2 + v_4$, as we considered in §1.3:

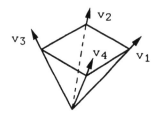

There are three obvious ways to resolve the singularity by subdividing:

Δ_1 Draw the plane through v_1 and v_3 (take $v = v_1$ or v_3);

Δ_2 Draw the plane through v_2 and v_4 (take $v = v_2$ or v_4);

Δ_3 Add a line through $v = v_1 + v_3 = v_2 + v_4$.

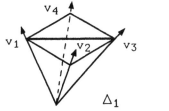
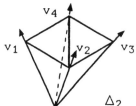
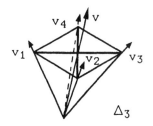

The first two of these replace σ by two simplices, the third by four. Since the third refines each of the first two, the corresponding resolution maps to each of them:

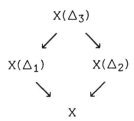

This description of two different minimal resolutions is well known for this cone over a quadric surface: each of $X(\Delta_1) \to X$ and $X(\Delta_2) \to X$ has fiber $\mathbb{P}^1$ over the singular point, corresponding to the two rulings of the quadric, while $X(\Delta_3) \to X$ has fiber $\mathbb{P}^1 \times \mathbb{P}^1$, the quadric itself. The transformation from $X(\Delta_1)$ to $X(\Delta_2)$ is an example of a "flop", which is a basic transformation in higher-dimensional birational geometry. In fact, toric varieties have provided useful models for Mori's program.[25]

Let us consider a global example. Let Δ be fan in $\mathbb{R}^3$ over the faces of the cube with vertices at $(\pm 1, \pm 1, \pm 1)$; take N to be the sublattice of $\mathbb{Z}^3$ generated by the vertices of the cube, as in Example (3) at the end of Chapter 1. The six singularities of $X(\Delta)$ can be resolved by doing any of the above subdivisions to each of the face of the cube. The following is a particularly pleasant way to do it:

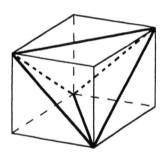

These added lines determine a tetrahedron; the fan $\overline{\Delta}$ of cones over faces of this tetrahedron determines the toric variety $X(\overline{\Delta}) = \mathbb{P}^3$. Since $\widetilde{\Delta}$ is also a refinement of $\overline{\Delta}$, we have morphisms

$$X(\Delta) \leftarrow X(\widetilde{\Delta}) \to X(\overline{\Delta}) = \mathbb{P}^3 .$$

Exercise. Show that the morphism $X(\widetilde{\Delta}) \to X(\overline{\Delta})$ is the blow-up of $\mathbb{P}^3$ along the four fixed points of the torus. In $X(\widetilde{\Delta})$ the proper transforms of the six lines joining these points are disjoint. Show that the morphism $X(\widetilde{\Delta}) \to X(\Delta)$ contracts these lines to the six singular points of $X(\Delta)$.

CHAPTER 3

ORBITS, TOPOLOGY, AND LINE BUNDLES

3.1 Orbits

As with any set on which a group acts, a toric variety $X = X(\Delta)$ is a disjoint union of its orbits by the action of the torus $T = T_N$. We will see that there is one such orbit O_τ for each cone τ in Δ; it is the orbit containing the distinguished point x_τ that was described in §2.1. Moreover,

$$O_\tau \cong (\mathbb{C}^*)^{n-k} \qquad \text{if} \quad \dim(\tau) = k .$$

If τ is n-dimensional, then O_τ is the point x_τ. If $\tau = \{0\}$, then $O_\tau = T_N$. We will see that O_τ is an open subvariety of its closure, which is denoted $V(\tau)$. The variety $V(\tau)$ is a closed subvariety of X that is again a toric variety. In fact, $V(\tau)$ will be a disjoint union of those orbits O_γ for which γ contains τ as a face.

Before working this out, let us look at the simplest example: $T = (\mathbb{C}^*)^n$ acting as usual on $X = \mathbb{C}^n$. The orbits are the sets

$$\{(z_1, \ldots, z_n) \in \mathbb{C}^n : z_i = 0 \text{ for } i \in I, \ z_i \neq 0 \text{ for } i \notin I\} ,$$

as I ranges over all subsets of $\{1, \ldots, n\}$. This is the orbit containing x_τ, where τ is generated by the basic vectors e_i for $i \in I$. All of the above assertions are evident in this example. Note also that

$$O_\tau = \text{Hom}(\tau^\perp \cap M, \mathbb{C}^*) ,$$

which is a formula that will be true generally as well. The general case of a nonsingular affine variety is obtained by crossing this example with a torus $(\mathbb{C}^*)^\ell$.

For a compact example, consider the projective space $\mathbb{P}^n$ corresponding to the fan of cones generated by proper subsets of $\{v_0, \ldots, v_n\}$, where the vectors generate the lattice and add to zero. If τ is generated by the subset $\{v_i : i \in I\}$, then $V(\tau)$ is the intersection of the hyperplanes $z_i = 0$ for $i \in I$, and O_τ consists of the points of $V(\tau)$ whose other coordinates are nonzero.

In the general case we will first describe the orbits O_τ and their closures $V(\tau)$ abstractly, and then show how to embed them in $X(\Delta)$. For each τ we defined N_τ to be the sublattice of N generated (as a group) by $\tau \cap N$, and

$$N(\tau) = N/N_\tau, \quad M(\tau) = \tau^\perp \cap M$$

the quotient lattice and its dual. Define O_τ to be the torus corresponding to these lattices:

$$O_\tau = T_{N(\tau)} = \text{Hom}(M(\tau), \mathbb{C}^*) = \text{Spec}(\mathbb{C}[M(\tau)]) = N(\tau) \otimes_\mathbb{Z} \mathbb{C}^* .$$

This is a torus of dimension $n-k$, where $k = \dim(\tau)$, on which T_N acts transitively via the projection $T_N \to T_{N(\tau)}$.

The *star* of a cone τ can be defined abstractly as the set of cones σ in Δ that contain τ as a face. Such cones σ are determined by their images in $N(\tau)$, i.e. by

$$\overline{\sigma} = \sigma + (N_\tau)_\mathbb{R} / (N_\tau)_\mathbb{R} \subset N_\mathbb{R}/(N_\tau)_\mathbb{R} = N(\tau)_\mathbb{R} .$$

These cones $\{\overline{\sigma} : \tau \prec \sigma\}$ form a fan in $N(\tau)$, and we denote this fan by $\text{Star}(\tau)$. (We think of $\text{Star}(\tau)$ as the cones containing τ, but realized as a fan in the quotient lattice $N(\tau)$.)

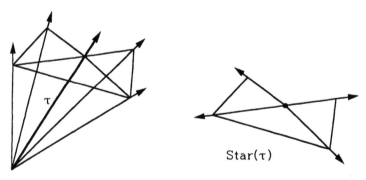

$\text{Star}(\tau)$

Set

$$V(\tau) = X(\text{Star}(\tau)) ,$$

the corresponding $(n-k)$-dimensional toric variety. Note that the torus embedding $O_\tau = T_{N(\tau)} \subset V(\tau)$ corresponds to the cone $\{0\} = \overline{\tau}$ in

$N(\tau)$.

   This toric variety $V(\tau)$ has an affine open covering $\{U_\sigma(\tau)\}$, as σ varies over all cones in Δ that contain τ as a face:

$$U_\sigma(\tau) = \mathrm{Spec}(\mathbb{C}[\overline{\sigma}^\vee \cap M(\tau)]) = \mathrm{Spec}(\mathbb{C}[\sigma^\vee \cap \tau^\perp \cap M]) .$$

Note that $\sigma^\vee \cap \tau^\perp$ is a face of $\sigma^\vee$ (the face corresponding to τ by duality). For $\sigma = \tau$, $U_\tau(\tau) = O_\tau$.

   To embed $V(\tau)$ as a closed subvariety of $X(\Delta)$, we construct a closed embedding of $U_\sigma(\tau)$ in U_σ for each $\sigma \succ \tau$. Regarding the points as semigroup homomorphisms, the embedding

$$U_\sigma(\tau) = \mathrm{Hom}_{sg}(\sigma^\vee \cap \tau^\perp \cap M, \mathbb{C}) \hookrightarrow \mathrm{Hom}_{sg}(\sigma^\vee \cap M, \mathbb{C}) = U_\sigma$$

is given by *extension by zero*; again, the fact that $\sigma^\vee \cap \tau^\perp$ is a face of $\sigma^\vee$ implies that the extension by zero of a semigroup homomorphism is a semigroup homomorphism. The corresponding surjection of rings

$$\mathbb{C}[\sigma^\vee \cap M] \longrightarrow\!\!\!\!\rightarrow \mathbb{C}[\sigma^\vee \cap \tau^\perp \cap M] ,$$

is the obvious *projection*: it takes χ^u to χ^u if u is in $\sigma^\vee \cap \tau^\perp \cap M$, and it takes χ^u to 0 otherwise.

   These maps are compatible: if τ is a face of σ, and σ a face of σ', the diagram

$$\begin{array}{ccc} U_\sigma(\tau) & \hookrightarrow & U_{\sigma'}(\tau) \\ \downarrow & & \downarrow \\ U_\sigma & \hookrightarrow & U_{\sigma'} \end{array}$$

commutes, since it comes from the commutative diagram

$$\begin{array}{ccc} \mathrm{Hom}_{sg}(\sigma^\vee \cap \tau^\perp \cap M, \mathbb{C}) & \hookrightarrow & \mathrm{Hom}_{sg}(\sigma'^\vee \cap \tau^\perp \cap M, \mathbb{C}) \\ \downarrow & & \downarrow \\ \mathrm{Hom}_{sg}(\sigma^\vee \cap M, \mathbb{C}) & \hookrightarrow & \mathrm{Hom}_{sg}(\sigma'^\vee \cap M, \mathbb{C}) \end{array}$$

where the horizontal maps are restrictions and the vertical maps are extensions by zero. These maps therefore glue together to give a closed embedding

$$V(\tau) \hookrightarrow X(\Delta) .$$

   If τ is a face of τ', we have closed embeddings

$$V(\tau') \hookrightarrow V(\tau)$$

given, on the open set U_σ for $\sigma \in \text{Star}(\tau')$, by extension by zero from $\text{Hom}_{sg}(\sigma^\vee \cap \tau'^\perp \cap M, \mathbb{C})$ to $\text{Hom}_{sg}(\sigma^\vee \cap \tau^\perp \cap M, \mathbb{C})$; alternatively, regard $V(\tau)$ as a toric variety and apply the preceding construction. In summary, we have an *order-reversing* correspondence from cones τ in Δ to orbit closures $V(\tau)$ in $X(\Delta)$.

It follows from this description that the ideal of $V(\tau) \cap U_\sigma$ in A_σ is $\oplus \mathbb{C} \cdot \chi^u$, the sum over all u in S_σ such that $\langle u, v \rangle > 0$ for v in the relative interior of τ. For a nonsingular surface, this agrees with the description in §1.1 of the curve defined by an edge of a fan.

Exercise. For a cone σ, show that every M-graded prime ideal in A_σ has this form for some face τ of σ. Show that every T_N-invariant closed subscheme of U_σ is defined by a graded ideal in A_σ. (1)

Exercise. Show similarly that if $X(\Delta)$ is an affine variety, then Δ consists of all faces of some cone σ, so $X(\Delta) = U_\sigma$.

Consider the singular example $X = U_\sigma$, where σ is the cone in $\mathbb{Z}^2$ generated by $v_1 = 2e_1 - e_2$ and $v_2 = e_2$. We saw in §1.1 that U_σ is a cone over a conic, defined in $\mathbb{C}^3$ by an equation $V^2 = UW$. Then $V(\sigma)$ is the vertex of the cone, which is the origin in $\mathbb{C}^3$, and, if τ_1 and τ_2 are the edges through v_1 and v_2, then $V(\tau_1)$ is the line $U = V = 0$, and $V(\tau_2)$ is the line $V = W = 0$. The orbits O_{τ_i} are the complements of the origin in these lines, and $O_{\{0\}} = T_N$ is the complement of these lines in X.

Proposition. *There are the following relations among orbits* O_τ, *orbit closures* $V(\tau)$, *and the affine open sets* U_σ:

(a) $U_\sigma = \coprod_{\tau \prec \sigma} O_\tau$;

(b) $V(\tau) = \coprod_{\gamma \succ \tau} O_\gamma$;

(c) $O_\tau = V(\tau) \smallsetminus \bigcup_{\gamma \underset{\neq}{\succ} \tau} V(\gamma)$.

Proof. For (a), note that a point of U_σ is given by a semigroup homomorphism $x: \sigma^\vee \cap M \to \mathbb{C}$. This point is in $T_N = O_{\{0\}}$ exactly when x does not take on the value 0, for then it extends to a homomorphism from M to $\mathbb{C}^*$. In general,

$$x^{-1}(\mathbb{C}^*) = \sigma^\vee \cap \tau^\perp \cap M$$

for some face τ of σ. This follows from the characterization of faces given in an exercise in §1.2: the sum of two elements of $\sigma^\vee$ cannot be in $x^{-1}(\mathbb{C}^*)$ unless both are in $x^{-1}(\mathbb{C}^*)$. This means precisely that x corresponds to a point of $O_\tau \subset U_\sigma(\tau)$.

For (c), passing to $N(\tau)$, i.e., working in the toric variety $V(\tau)$, we may assume $\tau = \{0\}$, in which case we must show that

$$T_N = X(\Delta) \smallsetminus \bigcup_{\gamma \ne \{0\}} V(\gamma) .$$

By intersecting with open sets of the form U_σ, this follows from (a). Then (b) follows from (c) by induction on the dimension.

It follows from the proposition that $X(\Delta)$ is a disjoint union of the O_τ, which are the orbits of the T_N-action. In addition, an orbit O_τ is contained in the closure of O_σ exactly when σ is a face of τ. In particular, the closed orbits are exactly those O_σ for σ a maximal cone in Δ. It also follows that if Δ° is the fan obtained from a fan Δ by removing some maximal cones σ (while retaining their faces), then $X(\Delta^\circ)$ is obtained from $X(\Delta)$ by removing the closed orbits O_σ.

If $\Delta = \Delta_P$ is constructed from a polytope P as in §1.5, there is a cone σ_Q for each face Q of P, so an invariant subvariety $V_Q = V(\sigma_Q)$ for each face. Note that V_Q is contained in $V_{Q'}$ exactly when Q is a face of Q', and the (complex) dimension of V_Q is equal to the (real) dimension of Q.

If $X(\Delta)$ is nonsingular, it follows from the definitions that each orbit closure $V(\tau)$ is nonsingular. It is a good general exercise now to compute the orbits and orbit closures in all of the toric varieties we have seen, including those that are singular.

Exercise. Show that $V(\tau) \cap U_\sigma$ is the disjoint union of the orbits O_γ as γ varies over all cones such that $\tau \prec \gamma \prec \sigma$. In particular, $V(\tau)$ is disjoint from U_σ if τ is not a face of σ.

Exercise. Show that the distinguished point x_τ is the origin (identity) of the torus O_τ. Show that O_τ is the unique closed T_N-orbit in U_τ.

Exercise. Let $\varphi_*: X(\Delta') \to X(\Delta)$ be the map arising from a homomorphism of lattices mapping a fan Δ' in N' to a fan Δ in N, as in §1.4. If φ maps N' onto N, show that φ_* maps the orbit $O_{\tau'}$ of $X(\Delta')$ onto the orbit O_τ of $X(\Delta)$, where τ is the smallest cone of Δ that contains τ'. If $\varphi^{-1}(|\Delta|) = |\Delta'|$, deduce that φ_* maps $V(\tau')$ onto $V(\tau)$. [2]

Exercise. Any v in N determines a mapping $\lambda_v: \mathbb{C}^* \to T_N$, so an action of $\mathbb{C}^*$ on $X(\Delta)$. Show that the set of fixed points of this action is the union of those $V(\gamma)$ for which v is in N_γ.

3.2   Fundamental groups and Euler characteristics

First we look at the fundamental group $\pi_1(X(\Delta))$ of a toric variety. Base points will be omitted in the notation for fundamental groups; they may be taken to be the origins of the embedded tori. The main fact is that complete toric varieties are simply connected. In fact,

Proposition. Let Δ be a fan that contains an n-dimensional cone. Then $X(\Delta)$ is simply connected.

Proof. The first observation is that the inclusion $T_N \hookrightarrow X(\Delta)$ gives a surjection

$$\pi_1(T_N) \longrightarrow\!\!\!\!\!\longrightarrow \pi_1(X(\Delta)) \,.$$

This is a general fact for the inclusion of any open subvariety of a *normal* variety; the point is that a normal variety is locally irreducible as an analytic space, so that its universal covering space cannot be disconnected by throwing away the inverse image of a closed subvariety.[3]

Now for any torus T_N, there is a canonical isomorphism

$$N \xrightarrow{\;\approx\;} \pi_1(T_N) \,,$$

defined by taking $v \in N$ to the loop $S^1 \subset \mathbb{C}^* \to T_N$, where $\mathbb{C}^* \to T_N$ is the map λ_v defined in §2.3. If v is in $\sigma \cap N$ for some cone σ, the loop can be contracted in U_σ, since $\lim_{z \to 0} \lambda_v(z) = x_\sigma$ exists in U_σ; in fact, we have seen that λ_v extends to a map from $\mathbb{C}$ to U_σ. The contraction is given by

$$\lambda_{v,t}(z) = \begin{cases} \lambda_v(tz) & z \in S^1, \ 0 < t \le 1 \\ x_\sigma & z \in S^1, \ t = 0 \end{cases} .$$

If σ is n-dimensional, then $\sigma \cap N$ generates N as a group, so the fact that such loops are trivial in U_σ implies that all loops are trivial.

Corollary. *If* σ *is a* k-*dimensional cone, then* $\pi_1(U_\sigma) \cong \mathbb{Z}^{n-k}$.

Proof. This follows from the fact that $U_\sigma = U_{\sigma'} \times (\mathbb{C}^*)^{n-k}$, and $\pi_1(\mathbb{C}^*) = \pi_1(S^1) = \mathbb{Z}$.

More intrinsically, if σ' is the cone in the lattice N_σ generated by σ, the fibration $U_{\sigma'} \to U_\sigma \to T_{N(\sigma)}$ induces a canonical isomorphism

$$\pi_1(U_\sigma) \xrightarrow{\ \cong\ } \pi_1(T_{N(\sigma)}) = N(\sigma) .$$

Exercise. Let Δ be the fan in $\mathbb{R}^2$ consisting of three cones: the origin, and the two rays through $2e_1 + e_2$ and $e_1 + 2e_2$. Show that $\pi_1(X(\Delta)) \cong \mathbb{Z}/3\mathbb{Z}$.

In complete generality, if N' is the subgroup of N generated by all $\sigma \cap N$, as σ varies over Δ, then

$$\pi_1((X(\Delta)) = N/N' .$$

To see this, note that for each σ, $\pi_1(U_\sigma) = N/N_\sigma$. By the general van Kampen theorem,

$$\pi_1((X(\Delta))) = \pi_1(\bigcup U_\sigma) = \varinjlim \pi_1(U_\sigma) = \varinjlim N/N_\sigma = N/\Sigma N_\sigma = N/N' .$$

For affine toric varieties, a similar argument shows more:

Proposition. *If* σ *is an* n-*dimensional cone, then* U_σ *is contractible.*

Proof. We want to define a homotopy

$$H : U_\sigma \times [0,1] \rightarrow U_\sigma$$

between the retraction $r: U_\sigma \rightarrow x_\sigma$ and the identity map. Choose a lattice point v in the interior of σ. Regarding the points of U_σ as semigroup homomorphisms from S_σ to $\mathbb{C}$, define H by

$$H(x \times t)(u) = t^{\langle u,v \rangle} \cdot x(u) \quad \text{for} \quad t > 0 ,$$

and $H(x \times 0) = x_\sigma$. It is easy to see that $H(x \times t)$ is a semigroup homomorphism whenever x is. For $u = 0$, $H(x \times t)(u) = x(0) = 1$ for all t. For $u \in S_\sigma \setminus \{0\}$, $\langle u,v \rangle > 0$, so $H(x \times t)(u) \rightarrow 0$ as $t \rightarrow 0$. It follows that $H(\varphi \times t)$ approaches x_σ as $t \rightarrow 0$, and, since generators of S_σ determine an embedding of U_σ in some $\mathbb{C}^m$, the resulting mapping is continuous.

Corollary. *If* $\dim(\sigma) = k$, *then* $O_\sigma \subset U_\sigma$ *is a deformation retract.*

Proof. One can use the same proof as in the proposition, or an isomorphism $U_\sigma = U_{\sigma'} \times O_\sigma$.

Corollary. *There is a canonical isomorphism* $H^i(U_\sigma;\mathbb{Z}) \cong \wedge^i(M(\sigma))$, *where* $M(\sigma) = \sigma^\perp \cap M$.

Proof. Since O_σ is the torus $T_{N(\sigma)}$, its cohomology is the exterior algebra on the dual $M(\sigma)$ of its first homology group $N(\sigma)$.

Knowing the cohomology of the basic open sets U_σ can give some information about the cohomology of $X(\Delta)$. When one has an open covering $X = U_1 \cup \ldots \cup U_\ell$ of a space, if all intersections of the open sets are simply connected, the cohomology of X can be computed as the Čech cohomology of the covering. In general one has a spectral sequence

$$E_1^{p,q} = \bigoplus_{i_0 < \ldots < i_p} H^q(U_{i_0} \cap \ldots \cap U_{i_p}) \Rightarrow H^{p+q}(X) . \quad (4)$$

Apply this to the covering by open sets $U_i = U_{\sigma_i}$, where the σ_i are the maximal cones of Δ:

$$E_1^{p,q} = \bigoplus_{i_0 < \ldots < i_p} \wedge^q M(\sigma_{i_0} \cap \ldots \cap \sigma_{i_p}) \Rightarrow H^{p+q}(X(\Delta)) .$$

In particular, this gives a calculation of the *topological Euler characteristic* $\chi(X(\Delta))$. For every cone τ that has dimension less than n, the alternating sum $\Sigma(-1)^q \operatorname{rank}(\wedge^q M(\tau))$ vanishes, while if the dimension is n, this alternating sum is one. Therefore

$$\chi(X(\Delta)) = \Sigma(-1)^{p+q} \operatorname{rank} E_1^{p,q} = \text{\# } n\text{-}dimensional\ cones\ in\ \Delta.$$

Assume all maximal cones in Δ are n-dimensional, as is the case if Δ is complete. Since each U_{σ_i} is contractible,

$$E_1^{0,q} = 0 \quad \text{for } q \geq 1 .$$

In addition, the complex $E_1^{\cdot,0}$ is

$$0 \rightarrow \bigoplus_i \mathbb{Z}_{\sigma_i} \rightarrow \bigoplus_{i<j} \mathbb{Z}_{\sigma_i \cap \sigma_j} \rightarrow \bigoplus_{i<j<k} \mathbb{Z}_{\sigma_i \cap \sigma_j \cap \sigma_k} \rightarrow \cdots ,$$

which is a Koszul complex (or the cochain complex of a simplex). This implies that

$$E_2^{p,0} = 0 \quad \text{for } p \geq 1 .$$

From the spectral sequence one therefore has

$$H^2(X(\Delta)) = E_\infty^{1,1} = E_2^{1,1} = \operatorname{Ker}(E_1^{1,1} \rightarrow E_1^{2,1})$$

$$= \operatorname{Ker}\left(\bigoplus_{i<j} M(\sigma_i \cap \sigma_j) \rightarrow \bigoplus_{i<j<k} M(\sigma_i \cap \sigma_j \cap \sigma_k) \right) .$$

Note that any element u of $M(\sigma) = \sigma^\perp \cap M$ gives a nowhere vanishing function χ^u on U_σ. An element in the above kernel gives a cocycle defining a line bundle on $X(\Delta)$.

Exercise. Verify that the torus T_N acts on such a line bundle, compatibly with its action on $X(\Delta)$. Verify that the above isomorphism with $H^2(X)$ takes the cocycle for a line bundle to the first Chern class of the line bundle.[5]

In particular, every element of $H^2(X)$ is the first Chern class of a line bundle of this type.

3.3   Divisors

On any variety X, a *Weil divisor* is a finite formal sum $\Sigma a_i V_i$ of irreducible closed subvarieties V_i of codimension one in X. A *Cartier divisor* D is given by the data of a covering of X by affine open sets U_α, and nonzero rational functions f_α called *local equations*, such that the ratios f_α / f_β are nowhere zero regular functions on $U_\alpha \cap U_\beta$. The *ideal sheaf* $\mathcal{O}(-D)$ of D is the subsheaf of the sheaf of rational functions generated by f_α on U_α; the inverse sheaf $\mathcal{O}(D)$ is the subsheaf of the sheaf of rational functions generated by $1/f_\alpha$ on U_α. Regarded as a line bundle, its transition functions from U_α to U_β are f_α / f_β. A Cartier divisor D determines a Weil divisor, denoted $[D]$, by

$$[D] = \sum_{\text{cod}(V,X)=1} \text{ord}_V(D) \cdot V ,$$

where $\text{ord}_V(D)$ is the order of vanishing of an equation for D in the local ring along the subvariety V. When X is normal these local rings are discrete valuation rings, so the notion of "order" is the naive one. For normal varieties, the map $D \mapsto [D]$ embeds the group of Cartier divisors in the group of Weil divisors. A nonzero rational function f determines a *principal* divisor $\text{div}(f)$ whose local equation in each open set is f.[6]

We will be primarily concerned with divisors on a toric variety $X = X(\Delta)$ that are mapped to themselves by the torus $T = T_N$. The irreducible subvarieties of codimension one that are T-stable correspond to edges (or rays) of the fan. Number the edges $\tau_1, \ldots, \tau_d$, and let v_i be the first lattice point met along the edge τ_i. These divisors are the orbit closures:

$$D_i = V(\tau_i) .$$

The T-*Weil divisors* are the sums $\Sigma a_i D_i$ for integers a_i.

We want to describe the Cartier divisors that are equivariant by T, which we call T-*Cartier divisors*. First, consider the affine case $X = U_\sigma$, with $\dim(\sigma) = n$. Let D be a divisor that is preserved by T, corresponding to a fractional ideal $I = \Gamma(X, \mathcal{O}(D))$. We claim that I is generated by a function χ^u for a unique $u \in \sigma^\vee \cap M$. This can be seen algebraically as follows. The fact that the divisor is T-invariant implies that I is graded by M, i.e., I is a direct sum of spaces $\mathbb{C} \cdot \chi^u$ over some set of u in M. Since I is principal at the distinguished point x_σ, $I/\mathfrak{m}I$ must be one-dimensional, where $\mathfrak{m}$ is the sum of all $\mathbb{C} \cdot \chi^u$ for $u \neq 0$. It follows that there is a unique u with $I = A_\sigma \cdot \chi^u$. Consequently a general T-Cartier divisor on U_σ has the form $\operatorname{div}(\chi^u)$ for some unique $u \in M$.

Lemma. *Let $u \in M$, and let v be the first lattice point along an edge τ. Then $\operatorname{ord}_{V(\tau)}(\operatorname{div}(\chi^u)) = \langle u, v \rangle$, so*

$$[\operatorname{div}(\chi^u)] = \Sigma_i \langle u, v_i \rangle D_i .$$

Proof. The order can be calculated on the open set $U_\tau \cong \mathbb{C} \times (\mathbb{C}^*)^{n-1}$, on which $V(\tau)$ corresponds to $\{0\} \times (\mathbb{C}^*)^{n-1}$. This reduces the calculation to the one-dimensional case, i.e., to the case where $N = \mathbb{Z}$, τ is generated by $v = 1$, and $u \in M = \mathbb{Z}$. Then χ^u is the monomial X^u, whose order of vanishing at the origin is u.

For example, look at the cone σ in $\mathbb{Z}^2$ generated by the vectors $v_1 = 2e_1 - e_2$ and $v_2 = e_2$, so U_σ is the cone over a conic, with two T-Weil divisors D_1 and D_2 (which are straight lines on the cone). If $u = (p, q) \in M = \mathbb{Z}^2$, then $\operatorname{div}(\chi^u) = (2p - q)D_1 + qD_2$. In particular, we see that D_1 and D_2 are not Cartier divisors, although $2D_1$ and $2D_2$ are.

Exercise. Let σ be the cone in $\mathbb{Z}^2$ generated by $v_1 = 2e_1 - e_2$ and $v_2 = -e_1 + 2e_2$, corresponding to divisors D_1 and D_2. Show that $a_1 D_1 + a_2 D_2$ is a Cartier divisor on U_σ if and only if $a_1 \equiv a_2 \pmod 3$.

Exercise. Let σ be the cone in $\mathbb{Z}^3$ generated by $v_1 = e_1$, $v_2 = e_2$, $v_3 = e_3$, and $v_4 = e_1 - e_2 + e_3$. Show that $a_1 D_1 + a_2 D_2 + a_3 D_3 + a_4 D_4$ is a Cartier divisor on U_σ if and only if $a_1 + a_3 = a_2 + a_4$.

Exercise. Take σ as in the preceding exercise, but replace the lattice by $N = \mathbb{Z} \cdot (1/2b)e_1 + \mathbb{Z} \cdot (1/b)e_2 + \mathbb{Z} \cdot (1/a)e_3 + \mathbb{Z} \cdot (1/2b)(e_1 + e_2 + e_3)$, where

a and b are positive integers, with a > 1 and gcd(a,2b) = 1. Find
the first lattice points along the four edges, and find which $\Sigma a_i D_i$ are
Cartier divisors on U_σ. In particular, show that no positive multiple of
ΣD_i is a Cartier divisor.

Exercise. Show that for each irreducible Weil divisor D_τ on an affine
toric variety U_σ, there is an effective Cartier divisor that contains D_τ
with multiplicity one.[7]

For a cone σ of dimension less than n, a T-Cartier divisor on
U_σ is of the form $\mathrm{div}(\chi^u)$ for some $u \in M$, but

$$\mathrm{div}(\chi^u) = \mathrm{div}(\chi^{u'}) \iff u - u' \in \sigma^\perp \cap M = M(\sigma).$$

To see this write $U_\sigma = U_{\sigma'} \times T_{N(\sigma)}$ as usual. Note that σ and σ'
have the "same" edges, so corresponding Weil divisors. Therefore
T-Cartier divisors on U_σ correspond to elements of $M/M(\sigma)$.

On a general toric variety $X(\Delta)$, it follows that a T-Cartier
divisor is defined by specifying an element $u(\sigma)$ in $M/M(\sigma)$ for each
cone σ in Δ, defining divisors $\mathrm{div}(\chi^{-u(\sigma)})$ on U_σ (the minus sign is
taken to conform to the literature). Equivalently, $\chi^{u(\sigma)}$ generates
the fractional ideal of $\mathcal{O}(D)$ on U_σ:

$$\Gamma(U_\sigma, \mathcal{O}(D)) = A_\sigma \cdot \chi^{u(\sigma)}.$$

These must agree on overlaps; i.e., when τ is a face of σ, $u(\sigma)$ must
map to $u(\tau)$ under the canonical map from $M/M(\sigma)$ to $M/M(\tau)$. In
short,

$$\begin{aligned}
\{\text{T-Cartier divisors}\} &= \varprojlim M/M(\sigma) \\
&= \mathrm{Ker}\Big(\bigoplus_i M/M(\sigma_i) \to \bigoplus_{i<j} M/M(\sigma_i \cap \sigma_j)\Big),
\end{aligned}$$

with σ_i the maximal cones, as in the preceding section.

Exercise. Show that a Weil divisor $\Sigma a_i D_i$ is a Cartier divisor if and
only if for each (maximal) cone σ there is a $u(\sigma) \in M$ such that for
all $v_i \in \sigma$, $\langle u(\sigma), v_i \rangle = -a_i$. [8]

Exercise. If Δ is simplicial, show that any Weil divisor D is a
$\mathbb{Q}$-Cartier divisor, i.e., some positive multiple of D is a Cartier divisor.

3.4   Line bundles

Let Pic(X) be the group of all line bundles, modulo isomorphism. For any irreducible variety X, the map $D \mapsto \mathcal{O}(D)$ gives a homomorphism from the group of Cartier divisors on X onto Pic(X), with kernel the group of principal divisors. Let $A_{n-1}(X)$ denote the group of all Weil divisors modulo the subgroup of divisors [div(f)] of rational functions. The map $D \mapsto [D]$ determines a homomorphism from Pic(X) to $A_{n-1}(X)$, which is an embedding when X is normal:

$$Pic(X) \hookrightarrow A_{n-1}(X) .$$

For X a toric variety, any $u \in M$ determines a principal Cartier divisor $div(\chi^u)$, giving a homomorphism from M to the group $Div_T X$ of T-Cartier divisors. The following proposition shows that Pic(X) can be computed by using only T-Cartier divisors and functions, and similarly for $A_{n-1}(X)$ with T-Weil divisors. In addition, it gives a recipe for calculating these groups for a complete toric variety X.

Proposition. *Let* $X = X(\Delta)$, *where* Δ *is a fan not contained in any proper subspace of* $N_\mathbb{R}$. *Then there is a commutative diagram with exact rows:*

$$
\begin{array}{ccccccccc}
0 & \to & M & \to & Div_T X & \to & Pic(X) & \to & 0 \\
 & & \| & & \downarrow & & \downarrow & & \\
0 & \to & M & \to & \overset{d}{\underset{i=1}{\bigoplus}} \mathbb{Z}\cdot D_i & \to & A_{n-1}(X) & \to & 0
\end{array}
$$

In particular, $rank(Pic(X)) \le rank(A_{n-1}(X)) = d - n$, *where* d *is the number of edges in the fan. In addition,* Pic(X) *is free abelian.*

Proof. First note that $X \smallsetminus \bigcup D_i = T_N$ is affine, with coordinate ring the unique factorization ring $\mathbb{C}[X_1, X_1^{-1}, \dots, X_n, X_n^{-1}]$, so all Cartier divisors and Weil divisors on T_N are principal. This gives an exact sequence[9]

$$A_{n-1}(\bigcup D_i) \;=\; \bigoplus_{i=1}^{d} \mathbb{Z}\cdot D_i \;\to\; A_{n-1}(X) \;\to\; A_{n-1}(T_N) \;=\; 0\;.$$

Next, note that if f is a rational function on X whose divisor is
T-invariant, then $f = \lambda\cdot\chi^u$ for some u $\in$ M and complex number
λ; this follows by restricting to the torus T_N. The lemma in the
preceding section and the fact that the v_i span $N_{\mathbb{R}}$ imply that u is
determined uniquely by f. This shows the exactness of the second row.
     If $\mathcal{L}$ is an algebraic line bundle, its restriction to T_N must
be trivial, so $\mathcal{L} = \mathcal{O}(D)$ for some Cartier divisor supported on $\bigcup D_i$;
indeed, writing $\mathcal{L} = \mathcal{O}(E)$ for some Cartier divisor E, take a rational
function whose divisor agrees with E on T_N, and set D = E - div(f).
Hence D is T-invariant as a Weil divisor, and therefore as a Cartier
divisor. The exactness of the upper row follows easily.
     Finally, the fact that Pic(X) is torsion free follows from the fact
that it is a subgroup of $\oplus M/M(\sigma)$, and each $M/M(\sigma)$ is a lattice, so
torsion free.

Corollary. *If all maximal cones of* Δ *are* n-*dimensional, then*
$Pic(X(\Delta)) \cong H^2(X(\Delta),\mathbb{Z})$.

Proof. We must map the group of T-Cartier divisors onto $H^2(X(\Delta),\mathbb{Z})$,
with kernel M. We have seen the isomorphisms:

$$(\text{T-Cartier divisors}) \;\cong\; \mathrm{Ker}\Big(\bigoplus_i M/M(\sigma_i) \;\to\; \bigoplus_{i<j} M/M(\sigma_i\cap\sigma_j)\Big)\,,$$

$$H^2(X(\Delta);\mathbb{Z}) \;\cong\; \mathrm{Ker}\Big(\bigoplus_{i<j} M(\sigma_i\cap\sigma_j) \;\to\; \bigoplus_{i<j<k} M(\sigma_i\cap\sigma_j\cap\sigma_k)\Big)\,.$$

An element $\oplus u_i$ in the first kernel is mapped to the sum $\oplus(u_j - u_i)$
in the second (noting that $M(\sigma_i) = 0$ since σ_i is n-dimensional). It
is an easy exercise to show that this map is surjective, with kernel
isomorphic to M.

Exercise. Give an example with maximal cones not n-dimensional
where the map from $Pic(X(\Delta))$ to $H^2(X(\Delta);\mathbb{Z})$ is not surjective.[10]

Exercise. Let Δ be the complete fan in $\mathbb{Z}^2$ with edges along
$v_1 = e_1$, $v_2 = -e_1 + me_2$, (m > 1), and $v_3 = -e_2$. Show that the Weil

divisor $a_1 D_1 + a_2 D_2 + a_3 D_3$ is a Cartier divisor on $X = X(\Delta)$ if and only if $a_1 + a_2 \equiv 0 \pmod m$. Show that

$$\text{Pic}(X) = \mathbb{Z} \cdot m \, D_2 \hookrightarrow A_1(X) = \mathbb{Z} \cdot D_2 .$$

Exercise. Let Δ be the complete fan in $\mathbb{Z}^2$ with edges along $v_1 = 2e_1 - e_2$, $v_2 = -e_1 + 2e_2$, and $v_3 = -e_1 - e_2$. Show that $a_1 D_1 + a_2 D_2 + a_3 D_3$ is a Cartier divisor on $X = X(\Delta)$ if and only if $a_1 \equiv a_2 \equiv a_3 \pmod 3$. Show that

$$\text{Pic}(X) = \mathbb{Z} \cdot 3 D_1 \cong \mathbb{Z} ,$$

$$A_1(X) = (\mathbb{Z} \cdot D_1 + \mathbb{Z} \cdot D_2)/\mathbb{Z} \cdot 3(D_1 - D_2) \cong \mathbb{Z} \oplus \mathbb{Z}/3\mathbb{Z} .$$

In particular, $A_{n-1}X$ can have torsion.

Exercise. Let $X = X(\Delta)$, where Δ is the fan in $\mathbb{Z}^3$ over the faces of the convex hull of the points e_1, e_2, e_3, $e_1 - e_2 + e_3$, and $-e_1 - e_3$. Show that $\text{Pic}(X) \cong \mathbb{Z}$, $A_2(X) \cong \mathbb{Z}^2$, and $A_2(X)/\text{Pic}(X) \cong \mathbb{Z}$.

Exercise. Let $X = X(\Delta)$, where Δ is the fan over the faces of the cube with vertices $(\pm 1, \pm 1, \pm 1)$, and N the sublattice of $\mathbb{Z}^3$ generated by the vertices. Show that $\text{Pic}(X) \cong \mathbb{Z}$, generated by a divisor that is the sum of the four irreducible divisors corresponding to the vertices of a face. Show that $A_2(X) \cong \mathbb{Z}^5$, and $A_2(X)/\text{Pic}(X) \cong \mathbb{Z}^4$.

In §1.5 we constructed a complete fan Δ that cannot be constructed as the faces over any subdivided polytope. In fact, the exercise proving that this fan is not a fan over the faces of a convex polytope actually showed that every line bundle on $X(\Delta)$ is trivial; i.e., $\text{Pic}(X(\Delta)) = 0$.

Exercise. For this fan, show that $A_2(X(\Delta)) \cong \mathbb{Z}^5 \oplus \mathbb{Z}/2\mathbb{Z}$.

Exercise. Let Δ be a fan such that all of its maximal cones are n-dimensional. Show that the following are equivalent:
  (i) Δ is simplicial;
  (ii)   Every Weil divisor on $X(\Delta)$ is a $\mathbb{Q}$-Cartier divisor;
  (iii) $\text{Pic}(X(\Delta)) \otimes \mathbb{Q} \to A_{n-1}(X(\Delta)) \otimes \mathbb{Q}$ is an isomorphism;
  (iv) $\text{rank}(\text{Pic}(X(\Delta))) = d - n.$ [11]

The data $\{u(\sigma) \in M/M(\sigma)\}$ for a Cartier divisor D defines a *continuous piecewise linear function* ψ_D on the support $|\Delta|$: the restriction of ψ_D to the cone σ is defined to be the linear function $u(\sigma)$; i.e.,

$$\psi_D(v) = \langle u(\sigma), v \rangle \quad \text{for } v \in \sigma .$$

The compatibility of the data makes this function well defined and continuous. Conversely, any continuous function on $|\Delta|$ that is linear and integral (i.e., given by an element of the lattice M) on each cone, comes from a unique T-Cartier divisor. If $D = \Sigma a_i D_i$, the function ψ_D is determined by the property that $\psi_D(v_i) = -a_i$; equivalently

$$[D] = \Sigma - \psi_D(v_i) D_i .$$

These functions behave nicely with respect to operations on divisors. For example, $\psi_{D+E} = \psi_D + \psi_E$, so $\psi_{mD} = m\,\psi_D$. Note that $\psi_{div(\chi^u)}$ is the linear function $-u$. If D and E are linearly equivalent divisors, it follows that ψ_D and ψ_E differ by a linear function u in M.

A T-Cartier divisor $D = \Sigma a_i D_i$ on $X(\Delta)$ also determines a rational convex polyhedron in $M_{\mathbb{R}}$ defined by

$$P_D = \{ u \in M_{\mathbb{R}} : \langle u, v_i \rangle \geq -a_i \text{ for all } i \}$$

$$= \{ u \in M_{\mathbb{R}} : u \geq \psi_D \text{ on } |\Delta| \} .$$

Lemma. *The global sections of the line bundle $\mathcal{O}(D)$ are*

$$\Gamma(X, \mathcal{O}(D)) = \bigoplus_{u \in P_D \cap M} \mathbb{C} \cdot \chi^u .$$

Proof. It follows from the lemma of the preceding section that

$$\Gamma(U_\sigma, \mathcal{O}(D)) = \bigoplus_{u \in P_D(\sigma) \cap M} \mathbb{C} \cdot \chi^u ,$$

where

$$P_D(\sigma) = \{ u \in M_{\mathbb{R}} : \langle u, v_i \rangle \geq -a_i \ \forall \ v_i \in \sigma \} .$$

These identifications are compatible with restrictions to smaller open sets. It follows that $\Gamma(X, \mathcal{O}(D)) = \cap \Gamma(U_\sigma, \mathcal{O}(D))$ is the corresponding direct sum over the intersection of the $P_D(\sigma) \cap M$, as required.

Exercise. Show that (i) $P_{mD} = mP_D$; (ii) $P_{D + div(\chi^u)} = P_D - u$; (iii) $P_D + P_E \subseteq P_{D+E}$.

When $|\Delta| = N_{\mathbb{R}}$, the variety $X(\Delta)$ is complete, and it is a general fact that cohomology groups of a coherent sheaf are finite dimensional on any complete variety. In the toric case this means that the polyhedron P_D is bounded:

Proposition. *If the cones in* Δ *span* $N_{\mathbb{R}}$ *as a cone, then* $\Gamma(X, \mathcal{O}(D))$ *is finite dimensional. In particular,* $P_D \cap M$ *is finite.*

Proof. If P_D were unbounded, by using the compactness of a sphere in $M_{\mathbb{R}}$ there would be a sequence of vectors u_i in P_D, and positive numbers t_i converging to zero, such that $t_i \cdot u_i$ converges to some nonzero vector u in $M_{\mathbb{R}}$. The fact that $\langle u_i, v_j \rangle \geq -a_j$ for all j implies that $\langle u, v_j \rangle \geq 0$ for all j. Since the v_i span $N_{\mathbb{R}}$, we must have $u = 0$, a contradiction.

The next proposition answers the question of when a line bundle is generated by its sections, i.e., when there are global sections of the bundle such that at every point at least one is nonzero. Recall that a real-valued function ψ on a vector space is (upper) *convex* if

$$\psi(t \cdot v + (1-t) \cdot w) \geq t\psi(v) + (1-t)\psi(w)$$

for all vectors v and w, and all $0 \leq t \leq 1$. For the simplest example of the toric variety $\mathbb{P}^1$ corresponding to the unique complete fan in $\mathbb{Z}$, if D_1 and D_2 are the divisors corresponding to the positive and negative edges, the divisor $D = a_1 D_1 + a_2 D_2$ corresponds to the function ψ_D on $\mathbb{R}$ defined by

$$\psi_D(x) = \begin{cases} -a_1 x & \text{if } x \geq 0 \\ -a_2 x & \text{if } x \leq 0 \end{cases}$$

This function is convex exactly when $a_1 + a_2$ is nonnegative. Since $\mathcal{O}(D) \cong \mathcal{O}(a_1 + a_2)$ on $\mathbb{P}^1$, this is exactly the criterion for $\mathcal{O}(D)$ to be generated by its sections.

We are concerned with continuous functions ψ on $N_{\mathbb{R}}$ such that the restriction $\psi|_\sigma$ to each cone is given by a linear function

$u(\sigma) \in M$. In this case convexity means that the graph of ψ lies
under the graph of $u(\sigma)$ for all n-dimensional cones σ, so that the
graph of ψ is "tent-shaped":

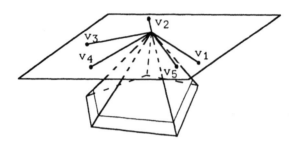

The convex function ψ is called *strictly convex* if the graph of ψ on
the complement of σ lies strictly under the graph of $u(\sigma)$, for all n-
dimensional cones σ; equivalently, for any n-dimensional cones σ
and σ', the linear functions $u(\sigma)$ and $u(\sigma')$ are different.

Proposition. *Assume all maximal cones in Δ are n-dimensional.
Let D be a T-Cartier divisor on $X(\Delta)$. Then $\mathcal{O}(D)$ is generated
by its sections if and only if ψ_D is convex.*

Proof. On any toric variety X, $\mathcal{O}(D)$ is generated by its sections if
and only if, for any cone σ, there is a $u(\sigma) \in M$ such that

(i) $\langle u(\sigma), v_i \rangle \geq -a_i$ for all i, and

(ii) $\langle u(\sigma), v_i \rangle = -a_i$ for those i for which $v_i \in \sigma$.

Indeed, (i) is the condition for $u(\sigma)$ to be in the polyhedron P_D that
determines global sections, and (ii) says that $\chi^{u(\sigma)}$ generates $\mathcal{O}(D)$ on
U_σ. The function ψ_D is determined by its restrictions to the n-
dimensional cones, where its values are given by (ii). The convexity of
ψ_D is then equivalent to (i).

If $\mathcal{O}(D)$ is generated by its sections, and all maximal cones of the
fan are n-dimensional, we can reconstruct D, or equivalently its
function ψ_D, from the polytope P_D:

$$\psi_D(v) = \min_{u \in P_D \cap M} \langle u, v \rangle = \min \langle u_i, v \rangle ,$$

where the u_i are the vertices of P_D.

Exercise. If $\mathcal{O}(D)$ and $\mathcal{O}(E)$ are generated by their sections, show that $P_{D+E} = P_D + P_E$.

Exercise. If $\mathcal{O}(D)$ is generated by its sections, and S is a subset of $P_D \cap M$, show that $\{\chi^u : u \in S\}$ generates $\mathcal{O}(D)$ if and only if S contains the vertices of P_D.

When $\mathcal{O}(D)$ is generated by its sections, choosing (and ordering) a basis $\{\chi^u : u \in P_D \cap M\}$ for the sections gives a mapping

$$\varphi = \varphi_D : X(\Delta) \rightarrow \mathbb{P}^{r-1}, \quad x \mapsto (\chi^{u_1}(x): \ldots : \chi^{u_r}(x)),$$

to projective space $\mathbb{P}^{r-1}$, $r = \text{Card}(P_D \cap M)$.

Lemma. *If* $|\Delta| = N_{\mathbb{R}}$, *the mapping* φ_D *is an embedding, i.e.,* D *is very ample, if and only if* ψ_D *is strictly convex and for every* n-*dimensional cone* σ *the semigroup* S_σ *is generated by*

$$\{u - u(\sigma) : u \in P_D \cap M\}.$$

Proof. $\Leftarrow$: Take corresponding homogeneous coordinates T_u on $\mathbb{P}^{r-1}$ indexed by the lattice points u in P_D. Let σ be an n-dimensional cone in the fan, and let $u(\sigma)$ be the corresponding element of $P_D \cap M$, so $\chi^{u(\sigma)}$ generates $\mathcal{O}(D)$ on U_σ. It follows easily from the strict convexity of ψ_D, that the inverse image by φ_D of the set $\mathbb{C}^{r-1} \subset \mathbb{P}^{r-1}$ where $T_{u(\sigma)} \neq 0$ is the open set U_σ. The restriction $U_\sigma \rightarrow \mathbb{C}^{r-1}$ is then given by the functions $\chi^{u-u(\sigma)}$, and the fact that they generate S_σ means that the corresponding map of rings is surjective, so the mapping is a closed embedding. The proof of the implication $\Rightarrow$ is similar.

Exercise. Let $X = X(\Delta)$, with Δ the fan over the faces of the cube with vertices $(\pm 1, \pm 1, \pm 1)$, N the lattice generated by the vertices. Let D be the sum of the divisors corresponding to the four vertices of the bottom of the cube. Show that P_D is the octahedron with vertices $(0,0,0)$, $(\frac{1}{2}, 0, \frac{1}{2})$, $(0, \frac{1}{2}, \frac{1}{2})$, $(0, -\frac{1}{2}, \frac{1}{2})$, $(-\frac{1}{2}, 0, \frac{1}{2})$, and $(0,0,1)$. Show that φ_D embeds X in $\mathbb{P}^5$, with image defined by the equations $T_0 T_5 = T_1 T_4 = T_2 T_3$. With $X(\overline{\Delta}) = \mathbb{P}^3$ as at the end of Chapter 2, show that this identifies X with the image of the rational map from $\mathbb{P}^3$ to $\mathbb{P}^5$ defined by the linear system of quadrics through the four T_N-fixed points: $(x_0: x_1: x_2: x_3) \mapsto (x_0x_1: x_0x_2: x_0x_3: x_1x_2: x_1x_3: x_2x_3)$. (12)

Proposition. *On a complete toric variety, a T-Cartier divisor D is ample, i.e., some positive multiple of D is very ample, if and only if its function ψ_D is strictly convex.*

Proof. Since $\psi_{mD} = m \psi_D$, the implication $\Rightarrow$ follows from the lemma. For the converse, note that replacing D by $m \cdot D$ replaces the polytope P_D by $m \cdot P_D = \{u \in M_\mathbb{R} : \langle u, v_i \rangle \geq -m \cdot a_i$ for all $i\}$. For any $u \in S_\sigma$ it follows from the fact that $\langle u(\sigma), v_i \rangle > -a_i$ if $v_i \notin \sigma$ that $u + m \cdot u(\sigma)$ is in $m \cdot P_D$ for large m. Since S_σ is a finitely generated semigroup, it follows that it is generated by elements $u - m \cdot u(\sigma)$ as u runs through $m \cdot P_D \cap M$, for m sufficiently large. By the lemma, it follows that mD is very ample.

Note in particular that for complete toric varieties, unlike for general complete varieties, *every ample line bundle is generated by its sections.*

Exercise. When D is ample, show that the polytope P_D is n-dimensional and the elements $u(\sigma)$, as σ varies over the n-dimensional cones of Δ, are exactly the vertices of P_D.

Exercise. For the toric variety $X = \mathbb{P}^n$ with its divisors $D_0, \ldots, D_n$, verify directly that $D = \Sigma a_i D_i$ is generated by its sections $\Leftrightarrow \psi_D$ is convex $\Leftrightarrow a_0 + \ldots + a_n \geq 0$, and that D is ample $\Leftrightarrow \psi_D$ is strictly convex $\Leftrightarrow a_0 + \ldots + a_n > 0$.

Exercise. For $X = \mathbb{F}_m$ the Hirzebruch surface, with edges generated by $v_1 = e_1$, $v_2 = e_2$, $v_3 = -e_1 + me_2$, $v_4 = -e_2$, show that $D = \Sigma a_i D_i$ is generated by its sections if and only if $a_2 + a_4 \geq 0$ and $a_1 + a_3 \geq m a_1$. Show that $\text{Pic}(X)$ is free on generators $A = D_1$ and $B = D_4$, and that $aA + bB$ is ample $\Leftrightarrow a$ and b are positive. Verify that A is a fiber of the mapping from $\mathbb{F}_m$ to $\mathbb{P}^1$, and B is the first Chern class of the universal quotient bundle of $\mathcal{O} \oplus \mathcal{O}(m)$ on $\mathbb{F}_m$.

Exercise. Show that any two-dimensional complete toric variety is projective. Show that any ample divisor on such a variety is very ample.[13]

Exercise. (Demazure) If $X(\Delta)$ is complete and nonsingular, show that a T-divisor is ample if and only if it is very ample.[14]

These ideas can be used to give simple examples of complete varieties that are not projective. Form the three-dimensional fan Δ whose edges in $\mathbb{Z}^3$ pass through $v_1 = -e_1$, $v_2 = -e_2$, $v_3 = -e_3$, $v_4 = e_1 + e_2 + e_3$, $v_5 = v_3 + v_4$, $v_6 = v_1 + v_4$, and $v_7 = v_2 + v_4$, and with cones through the faces of the triangulated tetrahedron shown:

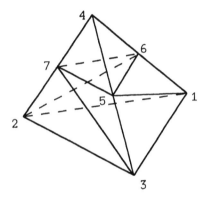

It is easy to verify that $X = X(\Delta)$ is complete and nonsingular. The claim is that X is not projective. Since $\text{Pic}(X)$ consists of line bundles of the form $\mathcal{O}(D)$ for D a T-Cartier divisor, it suffices to show that no function ψ_D can be strictly convex.

Exercise. If ψ were such a strictly convex function, show that

$$\psi(v_1) + \psi(v_5) > \psi(v_3) + \psi(v_6)$$
$$\psi(v_2) + \psi(v_6) > \psi(v_1) + \psi(v_7)$$
$$\psi(v_3) + \psi(v_7) > \psi(v_2) + \psi(v_5)$$

which add to a contradiction.

Exercise. (a) Describe the birational map from this variety X to $\mathbb{P}^3$ determined by this subdivision of the pyramid. In particular, show that the blowing up occurs over a plane triangle in $\mathbb{P}^3$.
(b) Show that the toric variety obtained by truncating the pyramid and omitting v_4 has a singular point of multiplicity 2.

Exercise. Prove the toric version of Chow's Lemma, that any complete toric variety can be dominated birationally by a projective toric variety.[15] Carry this out for the preceding example.

Any complete nonsingular toric variety, such as this example, has many nontrivial line bundles, since $\text{Pic}(X) \cong \mathbb{Z}^{d-n}$. We have seen an example of a singular complete toric variety that is even farther from being projective, having no nontrivial line bundles at all.[16]

We saw in §1.5 that a convex n-dimensional polytope P with vertices in M determines a fan Δ_P and a complete toric variety $X_P = X(\Delta_P)$. This variety comes equipped with a Cartier divisor $D = D_P$, whose corresponding convex function $\psi = \psi_D$ is defined by

$$\psi(v) = \min_{u \in P} \langle u,v \rangle = \min_{u \in P \cap M} \langle u,v \rangle = \min \langle u_i,v \rangle \,,$$

where the u_i are the vertices of P. Equivalently, if σ_i is the (maximal) cone corresponding to u_i, then χ^{u_i} generates $\mathcal{O}(D)$ on U_{σ_i}.

Exercise. Show that $P_D = P$, and verify that D is ample. Show conversely that if D is an ample T-Cartier divisor on a toric variety $X(\Delta)$, then the fan constructed from P_D is Δ.

Exercise. Given P as above, show that D_P is very ample if and only if for each vertex u of P, the semigroup generated by the vectors $u' - u$, as u' varies over $P \cap M$, is saturated.

Exercise. Let P be the polytope with vertices at $(0,0,0)$, $(1,0,0)$, $(0,0,1)$, $(1,1,0)$, and $(0,1,1)$ in $\mathbb{Z}^3$. Find the fan Δ_P, and show that $D = D_P$ is very ample, embedding X_P in $\mathbb{P}^4$ as the cone over a quadric surface, defined by an equation $T_1 T_4 = T_2 T_3$.

Exercise. Let $M = \mathbb{Z}^3$, P the polytope with vertices $(0,0,0)$, $(0,1,1)$, $(1,0,1)$, and $(1,1,0)$. Show that $D = D_P$ is ample but not very ample on $X = X_P$. Show in fact that $\varphi_D \colon X \to \mathbb{P}^3$ realizes X as the double cover of $\mathbb{P}^3$ branched along the four coordinate planes.

Exercise. Let P be an n-dimensional convex polytope with vertices in M, and assume that D_P is very ample, so we have $X_P \subset \mathbb{P}^{r-1}$, $r = \text{Card}(P \cap M)$. Let σ be the cone in $N \times \mathbb{Z}$ whose dual $\sigma^\vee$ is the cone over $P \times 1$ in $M \times \mathbb{Z}$. Identify the affine toric variety U_σ with the cone over X_P in $\mathbb{C}^r$. Deduce that $X_P \subset \mathbb{P}^{r-1}$ is *arithmetically normal* and *Cohen-Macaulay;* i.e., the homogeneous coordinate ring of X_P in $\mathbb{P}^{r-1}$ is normal and Cohen-Macaulay.[17]

Exercise. Let $P = \{(x_1, \ldots, x_n) \in \mathbb{R}^n : x_i \geq 0$ and $\Sigma x_i \leq m\}$. Show that the corresponding projective toric variety $X_P \subset \mathbb{P}^{r-1}$ is the m-fold Veronese embedding of $\mathbb{P}^n$ in $\mathbb{P}^{r-1}$, $r = \binom{n+m}{m}$. Show that the construction of the preceding exercise gives the cone over this embedding as the affine toric variety described in §2.2.

More generally, let P be the convex hull of any finite set in M. We call a complete fan Δ *compatible* with P if the function ψ_P defined by $\psi_P(v) = \min_{u \in P} \langle u, v \rangle$ is linear on each cone σ in Δ. Since ψ_P is convex, it determines a T-Cartier divisor $D = D_P$ on $X = X(\Delta)$ whose line bundle is generated by its sections. As before, these sections are linear combinations of the functions χ^u, as u varies over $P \cap M$.

Exercise. Show that the image of the corresponding morphism $\varphi_D \colon X \to \mathbb{P}^{r-1}$, is a variety of dimension k, where $k = \dim(P)$. [18]

3.5 Cohomology of line bundles

Let D be a T-Cartier divisor on a toric variety $X = X(\Delta)$, and let $\psi = \psi_D$ be the corresponding function on $|\Delta|$. We know that the sections of $\mathcal{O}(D)$ are a graded module: $H^0(X, \mathcal{O}(D)) = \oplus H^0(X, \mathcal{O}(D))_u$, where

$$H^0(X, \mathcal{O}(D))_u = \begin{cases} \mathbb{C} \cdot \chi^u & \text{if } u \in P_D \cap M \\ 0 & \text{otherwise} \end{cases},$$

with P_D the polyhedron $\{u \in M_\mathbb{R} : u \geq \psi$ on $|\Delta|\}$. This can be described in fancier words by defining a closed conical subset $Z(u)$ of $|\Delta|$ for each $u \in M$:

$$Z(u) = \{v \in |\Delta| : \langle u, v \rangle \geq \psi(v)\}.$$

Then u belongs to P_D exactly when $Z(u) = |\Delta|$, or equivalently, when the cohomology group $H^0(|\Delta| \smallsetminus Z(u))$ vanishes, where this H^0

denotes the 0th ordinary or sheaf cohomology of the topological space with complex coefficients. Equivalently, if

$$H^0_{Z(u)}(|\Delta|) = H^0(|\Delta|, |\Delta| \smallsetminus Z(u))$$

$$= \text{Ker}(H^0(|\Delta|) \to H^0(|\Delta| \smallsetminus Z(u)))$$

is the 0th *local cohomology* group (or relative group of the pair consisting of $|\Delta|$ and the complement of $Z(u)$), we have $u \in P_D$ exactly when $H^0_{Z(u)}(|\Delta|)$ is not zero. Therefore

$$H^0(X, \mathcal{O}(D)) = \oplus H^0(X, \mathcal{O}(D))_u \quad , \quad H^0(X, \mathcal{O}(D))_u = H^0_{Z(u)}(|\Delta|) .$$

This is the statement that generalizes to the higher sheaf cohomology groups $H^p(X, \mathcal{O}(D))$ and to the higher local cohomology groups $H^p_{Z(u)}(|\Delta|) = H^p(|\Delta|, |\Delta| \smallsetminus Z(u); \mathbb{C})$.

Proposition. *For all $p \geq 0$ there are canonical isomorphisms:*

$$H^p(X, \mathcal{O}(D)) \cong \oplus H^p(X, \mathcal{O}(D))_u \quad , \quad H^p(X, \mathcal{O}(D))_u \cong H^p_{Z(u)}(|\Delta|) .$$

These local cohomology groups are often easy to calculate. For example, if X is affine, so $|\Delta|$ is a cone and ψ is linear, then $|\Delta|$ and $|\Delta| \smallsetminus Z(u)$ are both convex, so all higher cohomology vanishes — which is one of the basic facts about general affine varieties. For toric varieties, a similar argument gives a stronger result than is usually true: *all higher cohomology groups of an ample line bundle on a complete toric variety vanish.* In fact, more is true:

Corollary. *If $|\Delta|$ is convex and $\mathcal{O}(D)$ is generated by its sections, then $H^p(X, \mathcal{O}(D)) = 0$ for all $p > 0$.*

Proof. Since ψ is a convex function, it follows that

$$|\Delta| \smallsetminus Z(u) = \{v \in |\Delta| : \langle u, v \rangle < \psi(v)\}$$

is a convex set, so both $|\Delta|$ and $|\Delta| \smallsetminus Z(u)$ are convex. This implies the vanishing of the corresponding cohomology groups.

It follows that for X complete and $\mathcal{O}(D)$ generated by its sections,

$$\chi(X,\mathcal{O}(D)) \;=\; \Sigma\,(-1)^p \dim H^p(X,\mathcal{O}(D))$$

$$=\; \dim H^0(X,\mathcal{O}(D)) \;=\; \mathrm{Card}(P_D \cap M)\;.$$

With Riemann-Roch formulas available to calculate the Euler characteristic, this gives an approach to counting lattice points in a convex polytope. We will come back to this in Chapter 5. Note in particular the formula for the arithmetic genus:

$$\chi(X,\mathcal{O}_X) \;=\; \dim H^0(X,\mathcal{O}_X) \;=\; 1\;.$$

Proof of the proposition. $H^p(\mathcal{O}(D))$ is the pth cohomology of the Čech complex $C^{\cdot}$, with

$$C^p \;=\; \bigoplus_{\sigma_0,\dots,\sigma_p} H^0(U_{\sigma_0} \cap \dots \cap U_{\sigma_p}, \mathcal{O}(D))$$

$$=\; \bigoplus_{u\,\epsilon\, M}\ \bigoplus_{\sigma_0,\dots,\sigma_p} H^0_{Z(u)\cap\sigma_0\cap \dots \cap\sigma_p}(\sigma_0 \cap \dots \cap\sigma_p)\;,$$

the sum over all cones $\sigma_0,\dots,\sigma_p$ in Δ. The boundary maps in the complex $C^{\cdot}$ preserve the M-grading, which gives its cohomology the grading. As before, we have $H^i_{Z(u)\cap|\tau|}(|\tau|) = 0$ for all cones τ and all u and all $i > 0$. The proposition then follows from a standard spectral sequence argument:

Lemma. *Let Z be a closed subspace of a space Y that is a union of a finite number of closed subspaces Y_j, and $\mathcal{F}$ a sheaf on Y such that $H^i_{Z\cap Y'}(Y',\mathcal{F}) = 0$ for all $i > 0$ and all $Y' = Y_{j_0} \cap \dots \cap Y_{j_p}$. Then*

$$H^i_Z(Y,\mathcal{F}) \;=\; H^i(C^{\cdot}((Y_j),\mathcal{F}))\;,$$

where $C^{\cdot}((Y_j),\mathcal{F})$ is the complex whose p^{th} term is

$$C^p((Y_j),\mathcal{F}) \;=\; \bigoplus_{j_0,\dots,j_p} \Gamma_{Z\cap Y_{j_0}\cap \dots \cap Y_{j_p}}(Y_{j_0} \cap \dots \cap Y_{j_p},\mathcal{F})\;.$$

Proof. Take an injective resolution $\mathcal{I}^{\cdot}$ of $\mathcal{F}$, and look at the double complex $C^{\cdot}((Y_j),\mathcal{I}^{\cdot})$. The hypothesis implies that the columns are resolutions of the complex $C^{\cdot}((Y_j),\mathcal{F})$. We claim that the rows are resolutions of the complex $\Gamma_Z(Y,\mathcal{I}^{\cdot})$. Then calculating the cohomology

of the total complex two ways (or appealing to the spectral sequence of
a double complex) gives the assertion of the lemma.

The exactness of the rows will follow from the fact that the $\mathcal{I}^q$
are injective. To see this, note that if $\mathcal{I}$ is injective, the sequence

$$0 \to \Gamma_{Z \cap W}(W, \mathcal{I}) \to \Gamma(W, \mathcal{I}) \to \Gamma(W \smallsetminus Z \cap W, \mathcal{I}) \to 0$$

is exact for any W. This reduces the assertion to the absolute case, i.e.,
to showing that

$$0 \to \Gamma(Y, \mathcal{I}) \to \oplus \Gamma(Y_j, \mathcal{I}) \to \oplus \Gamma(Y_{j_1} \cap Y_{j_2}, \mathcal{I}) \to \cdots$$

is exact. Since $\mathcal{I}$ is a direct summand of its sheaf $\tilde{\mathcal{I}}$ of arbitrary
("discontinuous") sections, we can replace $\mathcal{I}$ by $\tilde{\mathcal{I}}$. But then the
calculation is local at a point y in Y, and the cohomology is that of
the simplex $\{j : y \in Y_j\}$, with coefficients in the stalk $\mathcal{I}_y$.

Exercise. Let $X = \mathbb{P}^n$, a toric variety with its divisors $D_0, \ldots, D_n$ as
usual, and let $D = mD_0$, so $\mathcal{O}(D) \cong \mathcal{O}(m)$. Compute the cohomology as
follows. Show that ψ_D is zero on the cone generated by $v_1, \ldots, v_n$,
and ψ_D is $m \, e_i^*$ on the cone generated by $v_0, \ldots, \hat{v}_i, \ldots, v_n$.

(a)   For $m \geq 0$, verify that ψ_D is convex. Show directly that
$u \geq \psi_D$ exactly when $u = (m_1, \ldots, m_n)$ with $m_i \geq 0$ and $\Sigma m_i \leq m$;
the corresponding χ^u give a basis for $H^0(\mathbb{P}^n, \mathcal{O}(mD))$.

(b)   For $m < 0$, show that ψ_D is concave, so that the sets $Z(u)$
are convex and unequal to $N_{\mathbb{R}}$, and $H^i_{Z(u)}(N_{\mathbb{R}}) = 0$ unless $Z(u) = \{0\}$.
Use this to verify that $H^i(\mathbb{P}^n, \mathcal{O}(mD)) = 0$ for all $i \neq n$, and that
$H^n(\mathbb{P}^n, \mathcal{O}(mD)) \cong \oplus \mathbb{C} \cdot \chi^u$, the sum over those $u = (m_1, \ldots, m_n)$ with
$m_i < 0$ and $\Sigma m_i > m$.

Using the same techniques, we have the following important
result, which is also special to toric varieties:

Proposition. *Let Δ' be a refinement of Δ, giving a birational
proper map $f : X' = X(\Delta') \to X = X(\Delta)$. Then*

$$f_*(\mathcal{O}_{X'}) = \mathcal{O}_X \quad and \quad R^i f_*(\mathcal{O}_{X'}) = 0 \quad for \ all \ i > 0 \ .$$

In particular, taking X' to be a resolution of singularities, this says
that every toric variety X has *rational singularities*.

Proof. The assertion is local, so we can take Δ to be a cone σ together with all of its faces, so $X = U_\sigma$. The claims are that

(i) $\Gamma(X',\mathcal{O}_{X'}) = \Gamma(X,\mathcal{O}_X) = A_\sigma;$

(ii) $H^i(X',\mathcal{O}_{X'}) = 0$ for all $i > 0$.

The first is a general fact, since X is normal and f is birational; here it is obvious since both spaces of sections are $\oplus \mathbb{C}\cdot\chi^u$, the sum over $u \in \sigma^\vee \cap M$. The second follows from the fact that $\mathcal{O}_{X'}$ is generated by its sections, since the support $|\Delta'| = |\sigma|$ is convex.

CHAPTER 4

MOMENT MAPS AND THE TANGENT BUNDLE

4.1 The manifold with singular corners

Although we are working mainly with complex toric varieties, it is worth noticing that they are all defined naturally over the integers, simply by replacing $\mathbb{C}$ by $\mathbb{Z}$ in the algebras: $U_\sigma = \text{Spec}(\mathbb{Z}[\sigma^\vee \cap M])$. For a field K, the K-valued points of U_σ can be described as the semigroup homomorphisms

$$\text{Hom}_{sg}(\sigma^\vee \cap M, K),$$

where K is the multiplicative semigroup $K^* \cup \{0\}$. For example, for $K = \mathbb{R} \subset \mathbb{C}$, we have the *real* points of the toric variety.

In fact, the same holds when K is just a sub-semigroup of $\mathbb{C}$. The important case is the semigroup of nonnegative real numbers $\mathbb{R}_\geq = \mathbb{R}^+ \cup \{0\}$, which is a multiplicative sub-semigroup of $\mathbb{C}$. In this case there is a retraction given by the absolute value, $z \mapsto |z|$:

$$\mathbb{R}_\geq \subset \mathbb{C} \to \mathbb{R}_\geq.$$

For any cone σ, this determines a closed topological subspace

$$(U_\sigma)_\geq = \text{Hom}_{sg}(\sigma^\vee \cap M, \mathbb{R}_\geq) \subset U_\sigma = \text{Hom}_{sg}(\sigma^\vee \cap M, \mathbb{C})$$

together with a retraction $U_\sigma \to (U_\sigma)_\geq$. For any fan Δ, these fit together to form a closed subspace $X(\Delta)_\geq$ of $X(\Delta)$ together with a retraction

$$X(\Delta)_\geq \subset X(\Delta) \to X(\Delta)_\geq.$$

For example, if σ is generated by vectors $e_1, \ldots, e_k$ that form part of a basis for N, then $(U_\sigma)_\geq$ is isomorphic to a product of k copies of $\mathbb{R}_\geq$ and $n-k$ copies of $\mathbb{R}$. Thus if X is nonsingular, $X_\geq$ is a manifold with corners. When X is singular, the singularities of $X_\geq$ can be a little worse. For the toric variety $X = \mathbb{P}^n$, with its usual covering by affine open sets $U_i = U_{\sigma_i}$, $(U_i)_{\geq}$ consists of points $(t_0: \ldots :1: \ldots :t_n)$ with $t_i \geq 0$. Hence

$$\mathbb{P}^n_{\geq} = \mathbb{R}_{\geq}^{n+1} \smallsetminus \{0\} / \mathbb{R}^+$$

$$= \{(t_0, \ldots ,t_n) \in \mathbb{R}^{n+1} : \quad t_i \geq 0 \text{ and } t_0 + \ldots + t_n = 1\} ,$$

which is a standard n-simplex. The retraction from $\mathbb{P}^n$ to $\mathbb{P}^n_{\geq}$ is

$$(x_0 : \ldots : x_n) \mapsto \frac{1}{\Sigma |x_i|} (|x_0|, \ldots ,|x_n|) .$$

The fiber over a point $(t_0, \ldots ,t_n)$ is a compact torus of dimension equal to $\text{Card}\{i: t_i \neq 0\} - 1$.

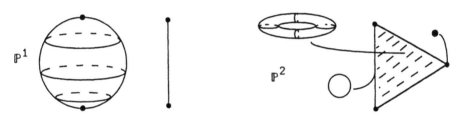

The algebraic torus T_N contains the *compact torus* S_N:

$$S_N = \text{Hom}(M,S^1) \subset \text{Hom}(M,\mathbb{C}^*) = T_N ,$$

where $S^1 = U(1)$ is the unit circle in $\mathbb{C}^*$. So S_N is a product of n circles. From the isomorphism of $\mathbb{C}^*$ with $S^1 \times \mathbb{R}^+ = S^1 \times \mathbb{R}$ (via the isomorphism of $\mathbb{R}^+$ with $\mathbb{R}$ given by the logarithm), we have

$$T_N = S_N \times \text{Hom}(M,\mathbb{R}^+) = S_N \times \text{Hom}(M,\mathbb{R}) = S_N \times N_{\mathbb{R}} ,$$

a product of a compact torus and a vector space.

Proposition. *The retraction* $X(\Delta) \to X(\Delta)_{\geq}$ *identifies* $X(\Delta)_{\geq}$ *with the quotient space of* $X(\Delta)$ *by the action of the compact torus* S_N.

Proof. Look at the action on the orbits O_τ:

$$(O_\tau)_{\geq} = X(\Delta)_{\geq} \cap O_\tau = \text{Hom}(\tau^\perp \cap M,\mathbb{R}^+)$$

$$= \text{Hom}(\tau^\perp \cap M,\mathbb{R}) = N(\tau)_{\mathbb{R}} .$$

From what we just saw, $S_{N(\tau)}$ acts faithfully on $O_\tau = T_{N(\tau)}$ with quotient space $(O_\tau)_{\geq} = N(\tau)_{\mathbb{R}}$. Since S_N acts on O_τ by way of its projection to $S_{N(\tau)}$, the conclusion follows.

Note that the fiber of $X \to X_2$ over $(O_\tau)_2$ can be identified with $S_{N(\tau)}$, which is a compact torus of dimension $n - \dim(\tau)$. The spaces $(O_\tau)_2$ fit together in X_2 in the same combinatorial way as the corresponding orbits O_τ in X. If one can get a good picture of the manifold with singular corners X_2, this can help in understanding how X is put together topologically.[1]

The manifold with corners X_2 can be described abstractly as a sort of "dual polyhedron" to Δ, at least if X is complete: X_2 has a vertex for each n-dimensional cone in Δ; two vertices are joined by an edge if the corresponding cones have a common (n-1)-face, and so on for smaller cones. If $\Delta = \Delta_P$ arises from a convex polytope in $M_{\mathbb{R}}$, then X_2 is homeomorphic to P. In the next section we will see an explicit realization of this homeomorphism. We list some other simple properties of this construction, leaving the verifications as exercises:

(1) If r is any positive number, the mapping $t \mapsto t^r$ determines an automorphism of $\mathbb{R}_2$. This determines an automorphism of the spaces $(U_\sigma)_2 = \mathrm{Hom}_{sg}(\sigma^\vee \cap M, \mathbb{R}_2)$, which fit together to determine a homeomorphism from X_2 to itself. If r is a positive integer, $z \mapsto z^r$ is an endomorphism of $\mathbb{C}$, which induces similarly an endomorphism of any toric variety X, compatible with the maps $X_2 \subset X \to X_2$.

(2) The quotient $T_N/S_N = N_{\mathbb{R}}$ acts on $X/S_N = X_2$, compatibly with the action of T_N on X. The inclusion $X_2 \subset X$ is equivariant with respect to the inclusion $N_{\mathbb{R}} = \mathrm{Hom}(M, \mathbb{R}^+) \subset \mathrm{Hom}(M, \mathbb{C}^*) = T_N$.

(3) There is a canonical mapping $S_N \times X_2 \to X$, which realizes X as a quotient space.

(4) For any cone τ, the inclusion $(O_\tau)_2 \subset (U_\tau)_2$ is a deformation retract.

Exercise. Use the Leray spectral sequence for the mapping $X \to X_2$ to reprove the result that the Euler characteristic of X is the number of n-dimensional cones.

4.2   Moment map

Moment maps occur frequently when Lie groups act on varieties.[2] Toric varieties provide a large class of concrete examples. In this section we construct these maps explicitly, and then sketch the relation to general moment maps.

Let P be a convex polytope in $M_{\mathbb{R}}$ with vertices in M, giving rise to a toric variety $X = X(\Delta_P)$ and a morphism $\varphi: X \to \mathbb{P}^{r-1}$ via the sections χ^u for $u \in P \cap M$ (see §3.4). Define a moment map

$$\mu: X \to M_{\mathbb{R}}$$

by

$$\mu(x) = \frac{1}{\Sigma|\chi^u(x)|} \sum_{u \in P \cap M} |\chi^u(x)|\, u .$$

Note that μ is S_N-invariant, since, for t in S_N and x in X, $|\chi^u(t \cdot x)| = |\chi^u(t)| \cdot |\chi^u(x)| = |\chi^u(x)|$. It follows that μ induces a map on the quotient space $X/S_N = X_{\geq}$:

$$\bar{\mu}: X_{\geq} \to M_{\mathbb{R}} .$$

Proposition. *The moment map defines a homeomorphism from* $X_{\geq}$ *onto the polytope* P.

In fact, one gets such a homeomorphism using any subset of the sections χ^u as long as P is the convex hull of the points, i.e., the subset contains the vertices of P.

Proof. Let Q be a face of P, and let σ be the corresponding cone of the fan. We claim that in fact $\bar{\mu}$ maps the subset $(O_\sigma)_{\geq}$ bijectively onto the relative interior of Q:

$$(O_\sigma)_{\geq} \xrightarrow{\;\approx\;} \mathrm{Int}(Q) ,$$

as a real analytic isomorphism.

Let $\rho_u(x) = |\chi^u(x)| / \Sigma|\chi^{u'}(x)|$, where the sum in the denominator is over all u' in $P \cap M$ (or in a subset containing the vertices of P). Therefore $0 \leq \rho_u(x) \leq 1$ and $\mu(x) = \Sigma\rho_u(x)u$. Note that a point x of $(O_\sigma)_{\geq}$ is in

$$\text{Hom}(\sigma^{\perp} \cap M, \mathbb{R}^{+}) \subset \text{Hom}_{sg}(\sigma^{\vee} \cap M, \mathbb{R}_{\geq}),$$

the inclusion by extending by zero outside $\sigma^{\perp}$. It follows that for x in $(O_{\sigma})_{2}$,

$$\rho_u(x) > 0 \quad \text{if } u \in Q; \quad \rho_u(x) = 0 \text{ if } u \notin Q.$$

Writing this out, one is reduced to proving the following assertion:

Lemma. *Let* V *be a finite-dimensional real vector space, and let* K *be the convex hull of a finite set of vectors* $u_1, \ldots, u_r$ *in the dual space* V^*. *Assume that* K *is not contained in a hyperplane. Let* $\varepsilon_1, \ldots, \varepsilon_r$ *be any positive numbers, and define* $\rho_i: V \to \mathbb{R}$ *by the formula*

$$\rho_i(x) = \varepsilon_i e^{u_i(x)} / (\varepsilon_1 e^{u_1(x)} + \ldots + \varepsilon_r e^{u_r(x)}).$$

Then the mapping $\mu: V \to V^*$, $\mu(x) = \rho_1(x)u_1 + \ldots + \rho_r(x)u_r$, *defines a real analytic isomorphism of* V *onto the interior of* K.

This is proved in the appendix to this section.

Exercise. Show that the vertices of the image of the moment map are the images of the points of X fixed by the action of the torus T_N.

The map from X_P to $\mathbb{P}^{r-1}$ is compatible with the actions of the torus T_N on X_P and the torus $T = (\mathbb{C}^*)^r$ on $\mathbb{P}^{r-1}$, with the map

$$T_N = \text{Hom}(M, \mathbb{C}^*) \to \text{Hom}(\mathbb{Z}^r, \mathbb{C}^*) = T$$

determined by the map $\mathbb{Z}^r \to M$ taking the basic vectors to the points of $P \cap M$. The action of the Lie group $S = (S^1)^r$ on $\mathbb{P}^{r-1}$ determines a *moment map*

$$\mathfrak{M} : \mathbb{P}^{r-1} \to \text{Lie}(S)^* = (\mathbb{R}^r)^* = \mathbb{R}^r.$$

If $x \in \mathbb{P}^{r-1}$ is represented by $v = (x_1, \ldots, x_r) \in \mathbb{C}^r$, then, up to a scalar factor, this moment map has the formula

$$\mathfrak{M}(x) = \frac{1}{\Sigma |x_i|^2} \sum_{i=1}^{r} |x_i|^2 e_i^*.$$

The following exercise shows that this agrees with a general construction of moment maps.

Exercise. Define $r_v : T/S \to \mathbb{R}$ by $r_v(t) = \frac{1}{2}\|t \cdot v\|^2$. The derivative of r_v at the origin e of the torus is a linear map

$$d_e(r_v) : \mathbb{R}^r = T_e(T/S) \to \mathbb{R}$$

i.e., $d_e(r_v)$ is in $(\mathbb{R}^r)^*$. Show that $\mathfrak{M}(x) = \|v\|^{-2} \cdot d_e(r_v)$.

· The composite

$$X_P \xrightarrow{\varphi} \mathbb{P}^{r-1} \xrightarrow{\mathfrak{M}} (\mathbb{R}^r)^* \to M_{\mathbb{R}}$$

is then a map from X_P to $M_{\mathbb{R}}$.

Exercise. Show that this composite takes x to $\mu(x^2)$, where $x \mapsto x^2$ is the map defined in (1) of the preceding section.

Appendix on convexity

The object is to prove the following elementary fact.

Proposition. *Let* $u_1, \ldots, u_r$ *be points in* $\mathbb{R}^n$, *not contained in any affine hyperplane, and let* K *be their convex hull. Let* $\varepsilon_1, \ldots, \varepsilon_r$ *be any positive real numbers, and define* $H : \mathbb{R}^n \to \mathbb{R}^n$ *by*

$$H(x) = \frac{1}{f(x)} \sum_{k=1}^{r} \varepsilon_k e^{(u_k, x)} u_k ,$$

where $f(x) = \varepsilon_1 e^{(u_1, x)} + \ldots + \varepsilon_r e^{(u_r, x)}$, *and* $(\,,\,)$ *is the usual inner product on* $\mathbb{R}^n$. *Then* H *defines a real analytic isomorphism of* $\mathbb{R}^n$ *onto the interior of* K.

We will deduce this from the following two related statements:

(A_n) *Let* $u_1, \ldots, u_r$ *be vectors in* $\mathbb{R}^n$ *that span* $\mathbb{R}^n$, *and let* C *be the cone (with vertex at the origin) that they span. Let* $\varepsilon_1, \ldots, \varepsilon_r$ *be any positive real numbers. Then the map* $F : \mathbb{R}^n \to \mathbb{R}^n$ *defined by*

$$F(x) = \sum_{k=1}^{r} \varepsilon_k e^{(u_k, x)} u_k$$

determines a real analytic isomorphism of $\mathbb{R}^n$ *onto the interior of* C.

($B_{n,m}$) *With* $C \subset \mathbb{R}^n$ *and* $F: \mathbb{R}^n \to \mathbb{R}^n$ *as in* (A_n), *let*
$\pi: \mathbb{R}^n \to \mathbb{R}^m$ *be a linear surjection. Then* $\pi \circ F$ *maps* $\mathbb{R}^n$ *onto*
the interior of $\pi(C)$, *and all of the fibers are connected manifolds*
isomorphic to $\mathbb{R}^{n-m}$.

The statement (A_1) is easy, since F is a mapping with positive
derivative, and it is easy to compute the limit of $F(x)$ as $x \to \pm\infty$.
More generally, for arbitrary n, and for $m \le n$, the matrix

$$\left(\frac{\partial F_i}{\partial x_j}(x) \right)_{1 \le i, j \le m}$$

is a positive definite symmetric matrix. Indeed, the value of this
quadratic form on a vector $t = (t_1, \ldots, t_m)$ is

$$\sum_{k=1}^{r} \varepsilon_k e^{(u_k, x)} \cdot (w_k, t)^2 ,$$

where w_k is the projection of u_k on the first m coordinates. In
particular, the Jacobian determinant of F is nowhere zero, so F is
a local isomorphism.

With this we can verify the implication (A_m) $\Rightarrow$ ($B_{n,m}$). After
changing coordinates, we may assume that $\pi: \mathbb{R}^n \to \mathbb{R}^m$ is the
projection to the first m coordinates. Let $\rho: \mathbb{R}^n \to \mathbb{R}^{n-m}$ be the
projection to the last $n-m$ coordinates. Let $G = \pi \circ F$, and for each
$y \in \mathbb{R}^{n-m}$, let $G_y: \rho^{-1}(y) = \mathbb{R}^m \to \mathbb{R}^m$ be the map induced by G.
Since $G_y(z) = \Sigma \varepsilon_k' e^{(w_k, z)} w_k$, with the ε_k' positive, and the
projections w_k span $\mathbb{R}^m$, the assertion (A_m) implies that each G_y
is a one-to-one map onto the interior of $\pi(C)$. It follows that for each
$q \in \text{Int}(\pi(C))$, the projection from $G^{-1}(q)$ to $\mathbb{R}^{n-m}$ induced by ρ is
one-to-one and onto. By the above Jacobian calculation, it is an
isomorphism of manifolds. This verifies ($B_{n,m}$).

We next verify that F is one-to-one. It suffices to show that its
restriction to a given line is one-to-one. After change of coordinates,
this line may be taken to be the line with x_i fixed for $2 \le i \le n$. If
$g(x_1)$ denotes the first component of $F(x)$, then (as in the case $m = 1$
above) the assumptions of case (A_1) are valid for $g: \mathbb{R} \to \mathbb{R}$. Since g
is one-to-one, the restriction of F to the line is one-to-one.

We show next that the image of F contains points arbitrarily close to any point on an edge of C. If several u_i lie on the same ray through the origin, they can be grouped in the formula for F, so one may assume no two u_i are on the same ray. Suppose the edge passes through u_1. Since the ray through u_1 is an edge, one may find a vector v so that $(u_1,v) = 0$ but $(u_j,v) < 0$ for $j > 1$. For any vector w, the limit of $F(\lambda v + w)$ as $\lambda \to +\infty$ is $e^{(u_1,w)}\varepsilon_1 u_1$. For appropriate choice of w, this limit can be placed arbitrarily along the edge.

It follows that once the image of F or $\pi \circ F$ is known to be convex, it must contain the whole interior of the cone.

Now we show the implication $(B_{n,n-1}) \Rightarrow (A_n)$. By what we have already proved, it suffices to show that the image of F is convex. Equivalently, we must show that the intersection of the image of F with any line is connected or empty. Any line is of the form $\pi^{-1}(q)$ for some projection π from $\mathbb{R}^n$ to $\mathbb{R}^{n-1}$, and some $q \in \mathbb{R}^{n-1}$. But

$$F(\mathbb{R}^n) \cap \pi^{-1}(q) = F((\pi \circ F)^{-1}(q)) ,$$

and $(\pi \circ F)^{-1}(q)$ is connected or empty by $(B_{n,n-1})$.

This, with an evident induction, completes the proofs of (A_n) and $(B_{n,m})$.

Finally, we show how the proposition follows from (A_{n+1}), by forming a cone over K in $\mathbb{R}^{n+1}$. Let

$$\tilde{u}_k = u_k \times 1 \in \mathbb{R}^n \times \mathbb{R} = \mathbb{R}^{n+1} ,$$

and define $\tilde{F}: \mathbb{R}^{n+1} \to \mathbb{R}^{n+1}$ by $\tilde{F}(x,t) = \Sigma \varepsilon_k e^{(u_k,x)} e^t \tilde{u}_k$. By (A_{n+1}) $\tilde{F}$ maps $\mathbb{R}^{n+1}$ isomorphically onto the interior of the cone $\tilde{C}$ generated by the $\tilde{u}_k$. The part mapping to $\text{Int}(K) \times 1$ consists of those (x,t) with $\Sigma \varepsilon_k e^{(u_k,x)} e^t = -1$, i.e., with $t = -\log f(x)$. So $H(x) = \tilde{F}(x,-\log f(x))$ maps $\mathbb{R}^n$ isomorphically onto the interior of K.

4.3 Differentials and the tangent bundle

Proposition. *If* X *is a nonsingular toric variety, and* $D_1, \ldots, D_d$ *are the irreducible T-divisors on* X, *then* $-\Sigma D_i$ *is a canonical divisor; i.e.,*

$$\Omega_X^n \;\cong\; \mathcal{O}_X\!\left(-\sum_{i=1}^{d} D_i\right).$$

Proof. If $e_1, \ldots, e_n$ are a basis for N, let $X_i = \chi^{e_i^*}$ be the coordinates corresponding to a dual basis for M, and set

$$\omega \;=\; \frac{dX_1}{X_1} \wedge \ldots \wedge \frac{dX_n}{X_n} \,,$$

a rational section of Ω_X^n. Another choice of basis for N gives the same differential form, up to multiplication by ± 1. We must show that the divisor of ω is $-\sum D_i$. On an open set U_σ, we may assume σ is generated by part of a basis, say $e_1, \ldots, e_k$, so we have

$$U_\sigma \;=\; \mathrm{Spec}\,(\mathbb{C}[X_1, \ldots, X_k, X_{k+1}, X_{k+1}^{-1}, \ldots, X_n, X_n^{-1}])$$

and

$$\omega \;=\; \frac{\pm 1}{X_1 \cdot \ldots \cdot X_n}\, dX_1 \wedge \ldots \wedge dX_n \,.$$

This shows that $\mathrm{div}(\omega)$ and $-\sum D_i$ have the same restriction to U_σ.

For example, let X be a nonsingular complete surface, with notation as in §2.5, so $(D_i \cdot D_i) = -a_i$. The canonical divisor $K = -\sum D_i$ has self-intersection number

$$(K \cdot K) \;=\; \sum(D_i \cdot D_i) + 2d \;=\; -\sum a_i + 2d \;=\; -(3d - 12) + 2d \;=\; 12 - d \,.$$

Since there are d two-dimensional cones, we also know that $\chi(X) = d$. Hence

$$\frac{(K \cdot K) + \chi(X)}{12} \;=\; \frac{(12 - d) + d}{12} \;=\; 1 \;=\; \chi(X, \mathcal{O}_X) \,,$$

which is *Noether's formula* for the surface X. We will use the Riemann-Roch formula in the next chapter to generalize this to higher dimensions.

We also need to know the whole sheaf Ω_X^1 of differential forms (the *cotangent bundle*) not just its top exterior power — at least up to exact sequences — so we can compute its Chern classes. For this we use the locally free sheaf $\Omega^1_X(\log D)$ of differentials with *logarithmic poles* along $D = \Sigma D_i$. At a point x in $D_1 \cap \ldots \cap D_k$, with x not in the other divisors, if $X_1, \ldots, X_n$ are local parameters such that $X_i = 0$ is a local equation for D_i, $1 \le i \le k$, then

$$\frac{dX_1}{X_1}, \frac{dX_2}{X_2}, \ldots, \frac{dX_k}{X_k}, dX_{k+1}, \ldots, dX_n$$

give a basis for $\Omega^1_X(\log D)$ at x. (3)

Proposition. (1) *There is an exact sequence of sheaves*

$$0 \to \Omega^1_X \to \Omega^1_X(\log D) \xrightarrow{d} \bigoplus_{i=1}^{d} \mathcal{O}_{D_i} \to 0 ,$$

where $\mathcal{O}_{D_i}$ is the sheaf of functions on D_i extended by zero to X.
(2) *The sheaf $\Omega^1_X(\log D)$ is trivial.*

Proof. For (2), consider the canonical map of sheaves

$$M \otimes_{\mathbb{Z}} \mathcal{O}_X \to \Omega^1_X(\log D)$$

that takes $u \in M$ to $d(\chi^u)/\chi^u$. To see that this map is an isomorphism, it suffices to look locally on affine open sets U_σ, where the assertion follows readily from the above description of $\Omega^1_X(\log D)$.

The second mapping in (1) is the *residue* mapping, which takes $\omega = \Sigma f_i dX_i/X_i$ to $\oplus f_i|_{D_i}$. The residue is zero precisely when each f_i is divisible by X_i, i.e., when ω is a section of Ω^1_X. (4)

4.4 Serre duality

For a vector bundle E on a nonsingular complete variety X, *Serre duality* gives isomorphisms

$$H^{n-i}(X, E^{\vee} \otimes \Omega^n_X) \cong H^i(X, E)^* .$$

If X is a toric variety and $E = \mathcal{O}(D)$ for a T-divisor D, this isomorphism respects the grading by M; with $\Omega^n_X = \mathcal{O}(-\Sigma D_i)$, it consists of isomorphisms

$$H^{n-i}(X,\mathcal{O}(-D-\Sigma D_i))_{-u} \;\cong\; (H^i(X,\mathcal{O}(D))_u)^*$$

for each $u \in M$. It is interesting to give a direct proof. Let $\psi = \psi_D$ be the piecewise linear function associated to D, and $\kappa = \psi_{-\Sigma D_i}$ the function for the canonical divisor $-\Sigma D_i$, so $\kappa(v_i) = 1$ for the lattice point v_i corresponding to each divisor D_i. By the description in §3.5,

$$H^i(X,\mathcal{O}(D))_u \;=\; H^i_Z(N_{\mathbb{R}})\,, \quad H^j(X,\mathcal{O}(-D-\Sigma D_i))_{-u} \;=\; H^j_{Z'}(N_{\mathbb{R}})\,,$$

where

$$Z \;=\; \{v \in N_{\mathbb{R}} : \psi(v) \le u(v)\}\,;$$

$$Z' \;=\; \{v \in N_{\mathbb{R}} : -\psi(v) + \kappa(v) \le -u(v)\}$$
$$\;=\; \{v \in N_{\mathbb{R}} : u(v) \le \psi(v) - \kappa(v)\}\,.$$

Serre duality amounts to isomorphisms

(SD) $$H^{n-i}_{Z'}(N_{\mathbb{R}}) \;\cong\; H^i_Z(N_{\mathbb{R}})^*\,.$$

The next three exercises outline a proof. Note that the set

$$S \;=\; \{v \in N_{\mathbb{R}} : \kappa(v) = 1\}$$

is the boundary of a polyhedral ball B, so is a deformation retract of the complement of $\{0\}$ in $N_{\mathbb{R}}$.

Exercise. If C is a nonempty closed cone in $N_{\mathbb{R}}$, show that there are canonical isomorphisms

$$H^i_C(N_{\mathbb{R}}) \;\cong\; H^i(N_{\mathbb{R}}, N_{\mathbb{R}} \smallsetminus C) \;\cong\; H^i(B, S \smallsetminus S \cap C)$$
$$\cong\; \tilde{H}^{i-1}(S \smallsetminus S \cap C) \;\cong\; \tilde{H}_{n-i-1}(S \cap C)\,,$$

the last by Alexander duality, where the $\sim$ denotes reduced cohomology and homology groups.

Exercise. Show that the embedding $S \cap Z \hookrightarrow S \smallsetminus (S \cap Z')$ is a deformation retract.[5]

Exercise. Prove (SD), first in the cases where $Z = N_{\mathbb{R}}$ and $Z' = \{0\}$, or $Z' = N_{\mathbb{R}}$ and $Z = \{0\}$; otherwise

$$H^{n-i}_{Z'}(N_{\mathbb{R}}) \;\cong\; \tilde{H}^{n-i-1}(S \smallsetminus S \cap Z') \;\cong\; \tilde{H}_{n-i-1}(S \smallsetminus S \cap Z')^*$$

$$\cong \tilde{H}_{n-i-1}(S \cap Z)^* \cong H_Z^i(N_{\mathbb{R}})^* . \text{ (6)}$$

Exercise. Suppose $X = X(\Delta)$ is a nonsingular toric variety, and $|\Delta|$ is a strongly convex cone in $N_{\mathbb{R}}$. Show that (i) $\Gamma(X, \Omega_X^n) = \oplus \mathbb{C} \cdot \chi^u$, the sum over all u in M that are positive on all nonzero vectors in $|\Delta|$; and (ii) $H^i(X, \Omega_X^n) = 0$ for $i > 0$. (7)

Grothendieck extended the Serre duality theorem to singular varieties. For this, the sheaf Ω_X^n must be replaced by a *dualizing complex* $\omega_X^{\cdot}$ in a derived category. When X is Cohen-Macaulay, however, this dualizing complex can be replaced by a single *dualizing sheaf* ω_X. For a vector bundle E on a complete n-dimensional Cohen-Macaulay variety X, Grothendieck duality gives isomorphisms

$$H^{n-i}(X, E^{\vee} \otimes \omega_X) \cong H^i(X, E)^* .$$

For a singular toric variety X, ΣD_i may not be a Cartier divisor (or even a $\mathbb{Q}$-Cartier divisor). Nevertheless, it defines a coherent sheaf $\mathcal{O}_X(-\Sigma D_i)$, whose local sections are rational functions with at least simple zeros along the divisors D_i. In fact, this is the dualizing sheaf:

Proposition. Let $\omega_X = \mathcal{O}_X(-\Sigma D_i)$.
(a) *If* $f: X' \to X$ *is a resolution of singularities obtained by refining the fan of* X, *then* $f_*(\Omega_{X'}^n) = \omega_X$ *and* $R^i f_*(\Omega_{X'}^n) = 0$ *for* $i > 0$.
(b) *If* X *is complete, and* L *is a line bundle on* X, *then*

$$H^{n-i}(X, L^{\vee} \otimes \omega_X) \cong H^i(X, L)^* .$$

Proof. Part (a) is local, so we may assume $X = U_\sigma$ for some cone σ. In this case, (a) is precisely what was proved in the preceding exercise. Then (b) follows; in fact, if E is any vector bundle on X, using Serre duality for $f^*(E)$ on X', we have

$$H^{n-i}(X, E^{\vee} \otimes \omega_X) \cong H^{n-i}(X, E^{\vee} \otimes Rf_*(\Omega_{X'}^n))$$
$$\cong H^{n-i}(X', f^*(E)^{\vee} \otimes \Omega_{X'}^n) \cong H^i(X', f^*(E))^* \cong H^i(X, E)^* ,$$

the last isomorphism using the last proposition in Chapter 3.

Exercise. Let $j: U \to X$ be the inclusion of the nonsingular locus U in X. Show that $j_*(\Omega_U^n) = \mathcal{O}_X(-\Sigma D_i)$. (8)

There is a pretty application of duality to lattice points in polytopes. If P is an n-dimensional polytope in $M_{\mathbb{R}}$ with vertices in M, we have seen that there is a complete toric variety X and an ample T-Cartier divisor D on X whose line bundle $\mathcal{O}(D)$ is generated by its sections, and these sections are linear combinations of χ^u for u in $P \cap M$. By refining the fan, one may take X to be nonsingular, if desired. Consider the exact sequence

$$0 \;\to\; \mathcal{O}(D - \Sigma D_i) \;\to\; \mathcal{O}(D) \;\to\; \mathcal{O}(D)|_{\Sigma D_i} \;\to\; 0 \;.$$

Exercise. (a) Show that $D - \Sigma D_i$ is generated by its sections, and these sections are

$$\bigoplus_{u \in \mathrm{Int}(P) \cap M} \mathbb{C} \cdot \chi^u \;.$$

(b) Deduce that

$$\chi(X, \mathcal{O}(D)|_{\Sigma D_i}) \;=\; h^0(X, \mathcal{O}(D)|_{\Sigma D_i}) \;=\; \mathrm{Card}(\partial P \cap M) \;,$$

where ∂P is the boundary of P.

Now by Serre-Grothendieck duality,

$$\chi(X, \mathcal{O}(D - \Sigma D_i)) \;=\; (-1)^n \chi(X, \mathcal{O}(-D)) \;.$$

Since the higher cohomology vanishes,

$$(-1)^n \chi(X, \mathcal{O}(-D)) \;=\; \mathrm{Card}(\mathrm{Int}(P) \cap M) \;.$$

It is a standard fact in algebraic geometry that for any Cartier divisor D on a complete variety X, the function $f \colon \mathbb{Z} \to \mathbb{Z}$,

$$f(\nu) \;=\; \chi(X, \mathcal{O}(\nu D)) \;,$$

is a polynomial in ν of degree at most $n = \dim(X)$, and this degree is n if D is ample.[9] In this case, for $\nu \geq 0$, $f(\nu) = \mathrm{Card}(\nu \cdot P \cap M)$. Putting this all together, we have the

Corollary. *If* P *is a convex n-dimensional polytope in* $M_{\mathbb{R}}$ *with vertices in* M, *there is a polynomial* f_P *of degree* n *such that*

$$\mathrm{Card}(\nu \cdot P \cap M) \;=\; f_P(\nu)$$

for all integers $\nu \geq 0$, *and*

$$\mathrm{Card}(\mathrm{Int}(\nu \cdot P) \cap M) = (-1)^n f_P(-\nu)$$

for all $\nu > 0$.

This formula, called the *inversion formula*, was conjectured by Ehrhart, and first proved by Macdonald in 1971. The above proof is from [Dani].

Exercise. Compute $f_P(\nu)$ for the polytope in $M = \mathbb{Z}^2$ with vertices at $(0,0)$, $(1,0)$, and $(1,b)$, for b a positive integer, and verify the inversion formula directly in this case.

This can be interpreted by means of the *adjunction formula*, which is an isomorphism $\Omega_X^n \otimes \mathcal{O}(D)|_D \cong \omega_D$, given by the residue. From the exact sequence

$$0 \to \Omega_X^n \to \Omega_X^n \otimes \mathcal{O}(D) \to \omega_D \to 0$$

and the long exact cohomology sequence, we see that

$$\Gamma(D, \omega_D) = \bigoplus_{u \in \mathrm{Int}(P) \cap M} \mathbb{C} \cdot \chi^u ,$$

$H^i(D, \omega_D) = 0$ for $0 < i < n-1$, and $\dim H^{n-1}(D, \omega_D) = 1$. Therefore

$$\chi(D, \mathcal{O}_D) = (-1)^{n-1} \chi(D, \omega_D) = 1 - \mathrm{Card}(\mathrm{Int}(P) \cap M) .$$

This means that the arithmetic genus of D is $\mathrm{Card}(\mathrm{Int}(P) \cap M)$.

For example, if P is the polytope in $\mathbb{Z}^2$ with vertices at $(0,0)$, $(d,0)$, and $(0,d)$, then $X = \mathbb{P}^2$, D is a curve of degree d, so its genus is $1 + 2 + \ldots + (d-2) = (d-1)(d-1)/2$. If P is a rectangle in $\mathbb{Z}^2$ with vertices at $(0,0)$, $(d,0)$, $(0,e)$, and (d,e), then $X = \mathbb{P}^1 \times \mathbb{P}^1$, D is a curve of bidegree (d,e), and the genus of D is $(d-1)(e-1)$. [10]

4.5 Betti numbers

For a smooth compact variety X, let $\beta_j = \mathrm{rank}(H^j(X))$ be its jth betti number. When $X = X(\Delta)$ is a toric variety, let d_k be the number of k-dimensional cones in Δ. In fact, these numbers determine each other:

Proposition. *If* $X = X(\Delta)$ *is a nonsingular projective toric variety,*
then $\beta_j = 0$ *if j is odd, and*

$$\beta_{2k} \;=\; \sum_{i=k}^{n} (-1)^{i-k} \binom{i}{k} d_{n-i} \; .$$

Set $h_k = \beta_{2k}$. If $P_X(t) = \Sigma \beta_j t^j$ is the Poincaré polynomial, then

$$P_X(t) \;=\; \Sigma h_k t^{2k} \;=\; \sum_{i=0}^{n} d_{n-i}(t^2-1)^i \;=\; \sum_{k=0}^{n} d_k(t^2-1)^{n-k} \; .$$

For example, for the topological Euler characteristic,

$$\chi(X) \;=\; \Sigma(-1)^j \beta_j \;=\; P_X(-1) \;=\; d_n \; ,$$

as we have seen.

Exercise. Invert the above formulas to express the d_k in terms of
the betti numbers:

$$d_k \;=\; \sum_{i=0}^{k} \binom{n-i}{n-k} h_{n-i} \; .$$

In fact, one can assign to *any* complex algebraic variety X (not
necessarily smooth, compact, or irreducible) a polynomial $P_X(t)$,
called its *virtual Poincaré polynomial*, with the properties:

    (1) $P_X(t) \;=\; \Sigma \operatorname{rank}(H^i(X)) t^i$ *if X is nonsingular and*
            projective (or complete);

    (2) $P_X(t) \;=\; P_Y(t) + P_U(t)$ *if Y is a closed algebraic subset of*
            X *and* $U = X \smallsetminus Y$.

For example, if U is the complement of r points in $\mathbb{P}^1$, then
$P_U(t) = t^2 + 1 - r$. (Note in particular that the coefficients can be
negative.) It is an easy exercise, using resolution of singularities and
induction on the dimension, to see that the polynomials are uniquely
determined by properties (1) and (2). Other properties follow easily
from (1) and (2):

    (3) *If X is a disjoint union of a finite number of locally closed*
subvarieties O(i), then $P_X(t) = \Sigma P_{O(i)}(t)$;

    (4) *If X = Y × Z, then* $P_X(t) = P_Y(t) \cdot P_Z(t)$.

Exercise. If $X \to Z$ is a fiber bundle with fiber Y that is locally trivial in the Zariski topology, show that $P_X(t) = P_Y(t) \cdot P_Z(t)$.

If X is nonsingular and complete, (1) says that $P_X(-1)$ is the Euler characteristic $\chi(X)$. With X, Y, and U as in (2), there is a long exact sequence

$$\ldots \to H_c^i U \to H_c^i X \to H_c^i Y \to H_c^{i+1} U \to H_c^{i+1} X \to \ldots ,$$

where H_c^* denotes cohomology with compact supports, and rational coefficients. It follows from this that $P_X(-1)$ must always be the *Euler characteristic with compact support*, i.e.,

$$P_X(-1) = \chi_c(X) = \sum (-1)^i \dim(H_c^i X) .$$

The existence of such polynomials follows from the existence of a mixed Hodge structure on these cohomology groups.[11] This gives a weight filtration on these vector spaces, compatible with the maps in the long exact sequence, such that the induced sequence of the mth graded pieces remains exact for all m:

$$\ldots \to \text{gr}_W^m(H_c^i U) \to \text{gr}_W^m(H_c^i X) \to \text{gr}_W^m(H_c^i Y) \to \text{gr}_W^m(H_c^{i+1} U) \to \ldots .$$

This means that the corresponding Euler characteristic

$$\chi_c^m(X) = \sum (-1)^i \dim(\text{gr}_W^m(H_c^i X))$$

is also additive in the sense of (2). If X is nonsingular and projective, then $\text{gr}_W^m(H_c^m X) = H_c^m X = H^m(X)$, so $\chi_c^m(X) = (-1)^m \dim(H^m(X))$. Hence we (must) set

$$P_X(t) = \sum_m (-1)^m \chi_c^m(X) t^m = \sum_{i,m} (-1)^{i+m} \dim(\text{gr}_W^m(H_c^i X)) t^m .$$

One need not know anything about the mixed Hodge structures or the weight filtration to use these virtual polynomials to calculate betti numbers; one has only to use the basic properties that determine them. For example, for a torus $T = (\mathbb{C}^*)^k$, we have $P_T(t) = (t^2 - 1)^k$. Hence if $X = X(\Delta)$ is an arbitrary toric variety, since it is a disjoint union of its orbits $O_\tau \cong T_{N(\tau)}$, by property (4) we have

$$(*) \quad P_{X(\Delta)}(t) = \sum P_{O_\tau}(t) = \sum d_{n-k}(t^2 - 1)^k$$

$$= \sum_{i=0}^{n} (\sum_{k=i}^{n} (-1)^{k-i}\binom{k}{i})d_{n-k})t^{2i},$$

where d_p denotes the number of cones of dimension p in Δ. This is true for any toric variety. In case $X(\Delta)$ is nonsingular and complete, however, this is the ordinary Poincaré polynomial by property (1), and this proves the proposition.

In fact, the proposition is also true when Δ is only simplicial and complete. For this one needs to know that $gr_W^m(H_c^m X) = H^m(X)$ also in this case. This follows from the combination of two facts: (i) the intersection (co)homology groups $IH^m(X)$ of an arbitrary compact variety have a mixed Hodge structure of pure weight m; (ii) if X is a V-manifold, then $H^m(X) = IH^m(X)$. [12]

A toric variety X is defined over the integers, so can be reduced modulo all primes. Since X is a disjoint union of orbits O_τ, and the number of $\mathbb{F}_q$-valued points of the torus $T_{N(\tau)} \cong (\mathbb{G}_m)^k$ is $(q-1)^k$, it follows that

$$\text{Card}(X(\mathbb{F}_q)) = \sum_{k=0}^{n} d_{n-k}(q-1)^k = \sum_{i=0}^{n} (\sum_{k=i}^{n} (-1)^{k-i}\binom{k}{i})d_{n-k})q^i.$$

When X is nonsingular and projective, Deligne's solution of the Weil conjectures implies that

$$\text{Card}(X(\mathbb{F}_{p^r})) = \sum_{j=0}^{2n} (-1)^j \sum_{i=0}^{\beta_j} \lambda_{ji}^r,$$

where the λ_{ji} are uniquely determined complex numbers with $|\lambda_{ji}| = p^{j/2}$.

Exercise. Use the preceding two formulas to give another proof of the proposition.

We will give a third proof of the proposition in Chapter 5.

Formula $(*)$ implies that for an arbitrary toric variety $X = X(\Delta)$, the Euler characteristic with compact support $\chi_c(X) = P_X(-1)$ is equal to the number of n-dimensional cones in Δ. We have seen earlier that the *ordinary* Euler characteristic $\chi(X)$ is also equal to the number of these cones. This raises the question of whether this is special to toric varieties, or is true for all varieties.

Exercise. Show that $\chi(X) = \chi_c(X)$ for every complex algebraic variety. Equivalently, show that $\chi(X) = \chi(Y) + \chi(U)$ whenever Y is a closed algebraic subset in a variety X with complement U. If N is a classical neighborhood of Y in X such that $\chi(Y) = \chi(N)$, show that $\chi(N \smallsetminus Y) = 0.$ [13]

CHAPTER 5

INTERSECTION THEORY

5.1 Chow groups

In this chapter we will work out some of the basic facts about intersections on a toric variety. On any variety X, the Chow group $A_k(X)$ is defined to be the free abelian group on the k-dimensional irreducible closed subvarieties of X, modulo the subgroup generated by the cycles of the form $[\mathrm{div}(f)]$, where f is a nonzero rational function on a (k+1)-dimensional subvariety of X. We have seen that on an arbitrary toric variety X the toric divisors generate the group $A_{n-1}(X)$ of Weil divisors modulo rational equivalence. The obvious generalization is valid as well:[1]

Proposition. *The Chow group $A_k(X)$ of an arbitrary toric variety $X = X(\Delta)$ is generated by the classes of the orbit closures $V(\sigma)$ of the cones σ of dimension n-k of Δ.*

Proof. Let $X_i \subset X$ be the union of all $V(\sigma)$ for all σ of dimension at least $n-i$. This gives a filtration $X = X_n \supset X_{n-1} \supset \ldots \supset X_{-1} = \varnothing$ by closed subschemes (say with reduced structure). The complement of X_{i-1} in X_i is the disjoint union of orbits O_σ, as σ varies over the cones of dimension $n-i$. From the exact sequence relating a closed subscheme and its complement, we have

$$A_k(X_{i-1}) \;\to\; A_k(X_i) \;\to\; \bigoplus_{\dim \sigma = n-i} A_k(O_\sigma) \;\to\; 0 .$$

Since a torus O_σ is an open subset of affine space $\mathbb{A}^i$, we see by the same principle that $A_i(O_\sigma) = \mathbb{Z}\cdot[O_\sigma]$ and $A_k(O_\sigma) = 0$ for $k \neq i$. Since the restriction from $A_k(X_i)$ to $A_k(O_\sigma)$ maps $[V(\sigma)]$ to $[O_\sigma]$, a simple induction shows that the classes $[V(\sigma)]$, $\dim(\sigma) = n-k$, generate $A_k(X_i)$.

For a Cartier divisor D on a variety X, the *support* of D is the union of the codimension one subvarieties W such that $\mathrm{ord}_W(D)$ is not zero. We say that D meets an irreducible subvariety V *properly* if V is not contained in the support of D. In this case one can define an *intersection cycle* $D \cdot V$ by restricting D to V (i.e., by restricting local defining equations), determining a Cartier divisor $D|_V$ on V, and taking the Weil divisor of this Cartier divisor: $D \cdot V = [D|_V]$. Let us work this out when X is a toric variety, $D = \Sigma a_i D_i$ is a T-Cartier divisor, and $V = V(\sigma)$. In this case, $D|_{V(\sigma)}$ will be a T-Cartier divisor on the toric variety $V(\sigma)$, so we will have

$$D \cdot V(\sigma) \ = \ \Sigma \, b_\gamma \, V(\gamma) \, ,$$

the sum over all cones γ containing σ with $\dim(\gamma) = \dim(\sigma) + 1$, and the b_γ are certain integers. To compute the multiplicity b_γ, suppose γ is spanned by σ and a finite set of minimal edge vectors v_i, $i \in I_\gamma$. Here is an example where I_γ has three vectors:

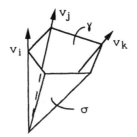

Let e be the generator of the one-dimensional lattice N_γ/N_σ such that the image of each v_i in N_γ/N_σ is a positive multiple of e, and let s_i be the integers such that v_i maps to $s_i \cdot e$ in N_γ/N_σ. Then b_γ is given by the formula

$$b_\gamma \ = \ \frac{a_i}{s_i} \qquad \text{for all } i \text{ in } I_\gamma \, .$$

To see this, for any cone γ containing σ let $u(\gamma) \in M/M(\gamma)$ be the linear function on γ corresponding to the divisor D. The assumption that $V(\sigma)$ is not contained in the support of D translates to the condition that $u(\gamma)$ vanishes on σ, which means that $u(\gamma)$ is in $M(\sigma)/M(\gamma)$. As γ varies over cones in the star of σ, these $u(\gamma)$ determine the divisor $D|_{V(\sigma)}$. In particular, when σ is a facet of γ, the multiplicity b_γ is $-\langle u(\gamma), e \rangle$. Therefore

$$a_i = -\langle u(\gamma),v_i \rangle = -\langle u(\gamma),s_i{\cdot}e \rangle = s_i(-\langle u(\gamma),e \rangle) = s_i{\cdot}b_\gamma \;,$$

as asserted.

When X is nonsingular, there is only one $i = i(\gamma)$ in I_γ, and $s_i = 1$, so $b_\gamma = a_{i(\gamma)}$ is the coefficient of $D_{i(\gamma)}$ in D. In this case, each D_k is a Cartier divisor, and

$$D_k{\cdot}V(\sigma) = \begin{cases} V(\gamma) & \text{if } \sigma \text{ and } v_k \text{ span a cone } \gamma \\ 0 & \text{if } \sigma \text{ and } v_k \text{ do not span a cone in } \Delta \end{cases}.$$

In fact, if $X(\Delta)$ is nonsingular, D_k and $V(\sigma)$ meet transversally in $V(\gamma)$ in the first case, and they are disjoint in the second case.

Exercise. If γ is simplicial, so there is one $i = i(\gamma)$ in I_γ, show that

$$b_\gamma = a_{i(\gamma)}{\cdot}\frac{\text{mult}(\sigma)}{\text{mult}(\gamma)} \;. \quad (2)$$

In general, if a subvariety V is contained in the support of a Cartier divisor D, the intersection $D{\cdot}V$ is defined only up to rational equivalence on V; i.e., $D{\cdot}V$ is in $A_{m-1}(V)$, where $m = \dim(V)$. This intersection class can be defined by finding a Cartier divisor E on V whose line bundle $\mathcal{O}_V(E)$ is isomorphic to the restriction of $\mathcal{O}_X(D)$ to V, and setting $D{\cdot}V$ to be the rational equivalence class of $[E]$. If f is a rational function on X such that V is not contained in the support of $D' = D + \text{div}(f)$, then $D{\cdot}V$ is represented by the cycle $D'{\cdot}V$ defined as above in the case of proper intersection.[3] For toric varieties, with D and $V(\sigma)$ as before, but with $V(\sigma)$ contained in the support of D, we can take some u in M so that $D' = D + \text{div}(\chi^u)$ meets $V(\sigma)$ properly. In fact, if $u(\sigma)$ is the linear function in $M/M(\sigma)$ defining D on σ, any u in M that maps to $u(\sigma)$ in $M/M(\sigma)$ will do. Then $D{\cdot}V(\sigma)$ is rationally equivalent to $D'{\cdot}V(\sigma)$, which is calculated as above. In the simplicial case, we get

$$D{\cdot}V(\sigma) \sim D'{\cdot}V(\sigma) = \sum (a_{i(\gamma)}\frac{\text{mult}(\sigma)}{\text{mult}(\gamma)} + \langle u,v_{i(\gamma)} \rangle) V(\gamma) \;,$$

where the sum is over all γ spanned by σ and one of the vectors $v_{i(\gamma)}$.

If V is a complete curve, the cycle or cycle class $D \cdot V$ has a well defined degree, which is denoted $(D \cdot V)$. The following exercise is the generalization to arbitrary dimension of the facts about intersecting divisors on a nonsingular surface that were seen in Chapters 1 and 2.

Exercise. Suppose an $(n-1)$-dimensional cone σ is the common face of two nonsingular n-dimensional cones γ' and γ''. Let $v_1, \ldots, v_{n-1}$ be the minimal lattice points on the edges of σ, and let v' and v'' be the minimal lattice points on the other edges of γ' and γ'' respectively.

(a) Show that there are (unique) integers $a_1, \ldots, a_{n-1}$ such that

$$v' + v'' = a_1 \cdot v_1 + a_2 \cdot v_2 + \ldots + a_{n-1} \cdot v_{n-1} \, .$$

(b) Show that, for $1 \le k \le n-1$,

$$D_k \cdot V(\sigma) \; \sim \; c_k' \, V(\gamma') + c_k'' \, V(\gamma'') \, ,$$

where c_k' and c_k'' are some integers whose sum is $-a_k$.

(c) Deduce that the intersection number is

$$(D_k \cdot V(\sigma)) \; = \; (D_1 \cdot D_2 \cdot \ldots \cdot D_k{}^2 \cdot \ldots \cdot D_{n-1}) \; = \; -a_k \, .$$

(d) Show that, in fact, $V(\sigma) \cong \mathbb{P}^1$, and $\mathcal{O}(D_k)|_{V(\sigma)} \cong \mathcal{O}(-a_k)$.

Exercise. If X is nonsingular and complete, show that a T-divisor D is ample if and only if $(D \cdot V(\sigma)) > 0$ for all cones σ of dimension $n-1$. [4]

On a nonsingular n-dimensional variety X, one sets $A^p(X) = A_{n-p}(X)$. There is an intersection product $A^p(X) \otimes A^q(X) \to A^{p+q}(X)$, making $A^*(X) = \oplus A^p(X)$ into a commutative, graded ring. For subvarieties V and W that meet properly, i.e., each component of their intersection has codimension equal to the sum of the codimensions of V and W, one can define an intersection cycle $V \cdot W$ by putting appropriate multiplicities in front of each component of the intersection; these multiplicities are one when the intersections are transversal. In general, one has only a rational equivalence class. [5] For a general toric variety $X(\Delta)$, if σ and τ are cones in Δ, then

$$V(\sigma) \cap V(\tau) \;=\; \begin{cases} V(\gamma) & \text{if } \sigma \text{ and } \tau \text{ span the cone } \gamma \\ \varnothing & \text{if } \sigma \text{ and } \tau \text{ do not span a cone in } \Delta \end{cases} .$$

This follows, even scheme-theoretically, from the description of these varieties in §3.1. This intersection is proper exactly when the dimension of γ is the sum of the dimensions of σ and τ. If Δ is nonsingular and the intersection is proper (or empty), then $V(\sigma)$ and $V(\tau)$ meet transversally in $V(\gamma)$ (or $\varnothing$), so $V(\sigma) \cdot V(\tau) = V(\gamma)$ (or 0). Alternatively, if σ has (minimal) generators $v_{i_1}, \dots, v_{i_k}$, then $V(\sigma)$ is the transversal intersection of $D_{i_1}, \dots, D_{i_k}$, and these formulas follow from the case of divisors considered above.

   If Δ is only simplicial, one has intersections of cycles or cycle classes only with rational coefficients. The Chow group

$$A^*(X)_{\mathbb{Q}} \;=\; \oplus A^P(X)_{\mathbb{Q}} \;=\; \oplus A_{n-p}(X) \otimes \mathbb{Q}$$

has the structure of a graded $\mathbb{Q}$-algebra. This is a general fact for any variety that is locally a quotient of a manifold by a finite group.[6] In case σ and τ span γ, with $\dim(\gamma) = \dim(\sigma) + \dim(\tau)$, then

$$V(\sigma) \cdot V(\tau) \;=\; \frac{\text{mult}(\sigma) \cdot \text{mult}(\tau)}{\text{mult}(\gamma)} \, V(\gamma) .$$

If σ and τ are contained in no cone of Δ, then $V(\sigma) \cdot V(\tau) = 0$. As in the nonsingular case, this follows from the description of the intersection of the divisors, using the fact that some positive multiple of each D_k is a Cartier divisor.

   One can often make calculations on singular varieties by resolving singularities. Any proper morphism $f: X' \to X$ determines a *push-forward* map $f_*: A_k(X') \to A_k(X)$, that takes the class of a variety V to the class of $\deg(V/f(V)) \cdot f(V)$ if $f(V)$ has the same dimension as V, and to 0 otherwise. In the toric case, if Δ' is a refinement of Δ, we have a proper birational map $f: X(\Delta') \to X(\Delta)$. We have seen that if σ' is a cone in Δ' that is contained in a cone σ of Δ of the same dimension, then f maps $V(\sigma')$ birationally onto $V(\sigma)$, so on cycles we have $f_*[V(\sigma')] = [V(\sigma)]$. If σ' is not contained in any cone of Δ of the same dimension, then $\dim(f(V(\sigma'))) < \dim(V(\sigma'))$, and $f_*[V(\sigma')] = 0$.

5.2 Cohomology of nonsingular varieties

Let $X = X(\Delta)$ be a nonsingular (or at least simplicial) complete toric variety. We have seen three proofs that

$$\chi(X) = m ,$$

where $m = d_n$ is the number of n-dimensional cones of X; in fact, we know formulas for the betti numbers of X in terms of the numbers of cones of various dimensions. On the other hand, we have potential generators for the homology groups, but with lots of relations, by using the orbit closures corresponding to cones of all dimensions. What we need is a way to associate to each n-dimensional cone σ a subcone τ, so that the resulting m varieties $V(\tau)$ give a basis for $H_*(X)$. [(7)] We will start with the assumption that X is nonsingular, and then point out the modifications needed when Δ is only simplicial.

For any ordering $\sigma_1, \ldots, \sigma_m$ of the top-dimensional cones, define a sequence of subcones $\tau_i \subset \sigma_i$, $1 \leq i \leq m$, by letting τ_i be the intersection of σ_i with all those σ_j that come after σ_i (i.e., with $j > i$) and that meet σ_i in a cone of dimension n-1. In other words, τ_i is the intersection of those walls of σ_i that are intersections with n-cones larger in the ordering. In particular, $\tau_1 = \{0\}$, and $\tau_m = \sigma_m$. The key assumption that will make this work is:

($*$)   If τ_i is contained in σ_j, then $i \leq j$.

For the following two-dimensional fan, the first two orderings satisfy ($*$), while the third does not:

Lemma. *If* X *is projective and* Δ *simplicial, then the n-dimensional cones of* Δ *can be ordered so that* ($*$) *holds.*

Proof. Take a very ample divisor, corresponding to a strictly convex function ψ, and for each n-dimensional cone σ let $u(\sigma) \in M$ be the element defining the restriction of ψ to σ. These $u(\sigma)$ are the vertices of a polytope P in $M_\mathbb{R}$ that describes the sections of the corresponding line bundle, and the idea is to choose a height function on M so that these vertices have different heights, and to order the σ by the value of the height function. That is, choose some $v \in N$ so that the m values $\langle u(\sigma), v \rangle$ are distinct, and order the σ's so that

$$\langle u(\sigma_1), v \rangle < \langle u(\sigma_2), v \rangle < \ldots < \langle u(\sigma_m), v \rangle .$$

Let $u_i = u(\sigma_i)$. Consider a fixed σ_i. Using the correspondence between cones in Δ and faces of P, τ_i is the cone corresponding to the smallest face of P that contains u_i and all edges of P that connect u_i to vertices u_j with $j > i$. The claim (∗) follows from the fact that this face contains no u_j with $j < i$.

Exercise. For the non-projective complete toric varieties considered in §3.4, find an ordering of the top-dimensional cones that verifies assumption (∗). Is there an example where (∗) is impossible to achieve?[8]

Theorem. *If (∗) holds and X is nonsingular, then the classes $[V(\tau_i)]$ form a basis for $A_*(X) \cong H_*(X; \mathbb{Z})$.*

The proof depends on a simple consequence of (∗):

Lemma. (a) *For each cone γ in Δ there is a unique $i = i(\gamma)$ such that $\tau_i \subset \gamma \subset \sigma_i$. In fact, $i(\gamma)$ is the smallest integer i such that σ_i contains γ.*

(b) *If γ is a face of γ', then $i(\gamma) \le i(\gamma')$.*

Proof. For the uniqueness in (a), if $\tau_i \subset \gamma \subset \sigma_i$ and $\tau_j \subset \gamma \subset \sigma_j$, then $\tau_i \subset \sigma_j$ and $\tau_j \subset \sigma_i$. Then (∗) implies that $i = j$. For the existence, given γ, let i be minimal such that $\gamma \subset \sigma_i$. Write γ as the intersection of some of the $(n-1)$-dimensional faces of σ_i; since these are all intersections of σ_i with some other n-dimensional cones, and by assumption these must come later in the sequence, γ must contain τ_i, as desired. Assertion (b) follows from the second statement in (a).

Now, for $1 \leq i \leq m$, set

$$Y_i = \bigcup_{\tau_i \subset \gamma \subset \sigma_i} O_\gamma = V(\tau_i) \cap U_{\sigma_i} ,$$

and

$$Z_i = Y_i \cup Y_{i+1} \cup \ldots \cup Y_m .$$

Lemma. (1) *Each Z_i is closed, $Z_1 = X$, and $Z_i \smallsetminus Z_{i+1} = Y_i$.*

(2) *If X is nonsingular, then $Y_i \cong \mathbb{C}^{n-k_i}$, where $k_i = \dim(\tau_i)$.*

Proof. It follows from (a) of the preceding lemma that X is a disjoint union of $Y_1, \ldots, Y_m$. Since the closure of O_γ is the union of all $O_{\gamma'}$ for γ' containing γ, the fact that Z_i is closed follows from (b) of the lemma, and the rest of assertion (1) is clear. Since $V(\tau_i) \cap U_{\sigma_i}$ is an affine open set in the toric variety $V(\tau_i)$ corresponding to a maximal $(n-k_i)$-dimensional cone in $N(\tau_i)$, (2) follows from what we saw in §2.1.

We can now prove the theorem, showing by descending induction on i that the canonical map $A_*(Z_i) \to H_*(Z_i)$ is an isomorphism, and that the classes of the closures $\overline{Y}_j = V(\tau_j)$, $j \geq i$, form a basis. We have a commutative diagram with exact rows:

$$A_p(Z_{i+1}) \to A_p(Z_i) \to A_p(Y_i) \to 0$$

$$\downarrow \qquad\qquad \downarrow \qquad\qquad \downarrow$$

$$\to H_{2p+1}(Y_i) \to H_{2p}(Z_{i+1}) \to H_{2p}(Z_i) \to H_{2p}(Y_i) \to H_{2p-1}(Z_{i+1}) \to$$

where the second row is the long exact sequence of *Borel-Moore* homology, i.e., homology with locally finite chains. Since Y_i is an affine space, $A_*(Y_i) \cong H_{2*}(Y_i) \cong \mathbb{Z}$, generated by the class of Y_i. By induction, the first vertical map is an isomorphism and $H_q(Z_{i+1}) = 0$ for q odd. A diagram chase shows that the middle vertical arrow is an isomorphism, and that $A_*(Z_i)$ is free on the classes $[V(\tau_j)]$, for $j \geq i$.

If Δ is only simplicial, all of the above argument is valid, provided rational coefficients are used in place of integers. The difference is that now each Y_i is a quotient $\mathbb{C}^{n-k_i}/G_i$ of affine space by a finite group, and for such a quotient, the Chow groups and Borel-Moore homology are the subspaces invariant by the group:

$$A_*(\mathbb{C}^r/G)_\mathbb{Q} = (A_*(\mathbb{C}^r)_\mathbb{Q})^G \, , \quad H_*(\mathbb{C}^r/G;\mathbb{Q}) = (H_*(\mathbb{C}^r;\mathbb{Q}))^G \, .$$

Theorem. *If* $(*)$ *holds and* Δ *is simplicial, then the classes* $[V(\tau_i)]$ *form a basis for the vector space* $A_*(X)_\mathbb{Q} \cong H_*(X;\mathbb{Q})$.

Corollary. *With the same assumptions,*

$$d_p = \sum_{k=0}^{p} \binom{n-k}{n-p} h_{n-k} \, ,$$

where d_p *is the number of* p-*dimensional cones in* Δ *and* h_i *is the rank of* $A_i(X)$ *(or* $H_{2i}(X)$*). Equivalently,*

$$h_k = \sum_{i=k}^{n} (-1)^{i-k} \binom{i}{k} d_{n-i} \, .$$

Proof. The dimension of $A_{n-k}(X)_\mathbb{Q}$ is the number of cones τ_i of dimension k. The number of p-dimensional cones γ with $\tau_i \subset \gamma \subset \sigma_i$ is $\binom{n-k}{n-p}$, and, by (a) of the second lemma, each cone occurs exactly once in this way.

It follows from the fact that X is locally a quotient of a manifold by a finite group that Poincaré duality is valid:

$$H^i(X;\mathbb{Q}) \otimes H^{2n-i}(X;\mathbb{Q}) \to H^{2n}(X;\mathbb{Q}) = \mathbb{Q}$$

is a perfect pairing. Since $H^i(X;\mathbb{Q})$ is dual to $H_i(X,\mathbb{Q})$, we deduce

Corollary 1. *For* Δ *simplicial,* $h_k = h_{n-k}$ *for all* $0 \le k \le n$.

In our situation we can see this explicitly as follows. With $\sigma_1, \ldots, \sigma_m$ and the corresponding $\tau_1, \ldots, \tau_m$ as above, let τ_i' be the intersection of σ_i with all σ_j such that j is *less* than i and $\dim(\sigma_j \cap \sigma_i) = n-1$.

Exercise. Show that $V(\tau_1'), \ldots, V(\tau_m')$ give a basis for $A_*(X)_\mathbb{Q}$. From the fact that $\dim(\tau_i') + \dim(\tau_i) = n$ for all i, deduce that $h_k = h_{n-k}$. Use the intersections of the varieties $V(\tau_i)$ and $V(\tau_j')$ to show directly that the intersection pairings $A_k(X)_{\mathbb{Q}} \otimes A_{n-k}(X)_{\mathbb{Q}} \to \mathbb{Q}$ are perfect pairings.

The fact that X is an orbifold (or V-manifold), i.e., locally a quotient of a manifold by a finite group, has another important consequence: both the cohomology and homology (with rational coefficients) coincide with the *intersection homology* $IH^*(X;\mathbb{Q})$. That is, the canonical maps

$$H^i(X;\mathbb{Q}) \;\to\; IH^i(X;\mathbb{Q}) \;\to\; H_{2n-i}(X;\mathbb{Q})$$

are isomorphisms. It has been established — first using Deligne's proof of the Weil conjectures, then by Saito's putting a pure Hodge structure on the intersection homology groups — that for any projective variety X, the intersection homology satisfies the *hard Lefschetz theorem*: if ω in $H^2(X)$ is the class of a hyperplane section, the maps

$$IH^{n-i}(X;\mathbb{Q}) \xrightarrow{\;\cup\, \omega^i\;} IH^{n+i}(X;\mathbb{Q})$$

are all isomorphisms.[9] This implies in particular that multiplication by ω is injective from IH^i to IH^{i+2} for $i < n$. For a V-manifold, this implies that the rank of $H^{2p-2}(X)$ is no more than the rank of $H^{2p}(X)$ for $2p \leq n$, which gives

Corollary 2. *For* Δ *simplicial,* $h_{p-1} \leq h_p$ *for* $1 \leq p \leq [\frac{n}{2}]$.

It would be interesting to prove hard Lefschetz for projective orbifolds without appealing to the deeper theorem for intersection homology. Even for toric varieties, proving this fact has been a challenge. If Δ is not simplicial, the homology and cohomology of $X(\Delta)$ can be much more complicated, and in fact need not be combinatorial invariants of the fan. However, the intersection homology of an arbitrary toric variety does have a such a description, which gives relations between the ranks of the intersection homology and the numbers of cones of each dimension. Then hard Lefschetz gives inequalities among these numbers.[10]

Exercise. Let $X = X(\Delta)$, where Δ is the cone over the faces of the cube with vertices at $(\pm 1, \pm 1, \pm 1)$, and lattice generated by the vertices. Show that $A_0(X) = H_0(X) = \mathbb{Z}$, $A_3(X) = H_6(X) = \mathbb{Z}$, and

$$A_1(X) = H_2(X) \cong \mathbb{Z}, \quad A_2(X) = H_4(X) \cong \mathbb{Z}^{\oplus 5}, \quad H_3(X) \cong \mathbb{Z}^{\oplus 2}. \quad (11)$$

Finally, we want to describe the intersection ring $A^*X = H^*X$ of a nonsingular projective toric variety $X = X(\Delta)$. Let $D_1, \ldots, D_d$ be the irreducible T-divisors, corresponding to the minimal lattice points $v_1, \ldots, v_d$ along the edges. By what we saw in the last section, if a cone σ is generated by $v_{i_1}, \ldots, v_{i_k}$, then $V(\sigma)$ is the transversal intersection of $D_{i_1}, \ldots, D_{i_k}$. Since the varieties $V(\sigma)$ span the cohomology, this means that A^*X is generated as a $\mathbb{Z}$-algebra by the classes of $D_1, \ldots, D_d$. We have seen that $D_{i_1} \cdot \ldots \cdot D_{i_k} = 0$ in A^kX if $v_{i_1}, \ldots, v_{i_k}$ do not generate a cone, since the intersection of the divisors is empty. In addition, if u is any element of M, the divisor of the rational function χ^u on X is $\sum \langle u, v_i \rangle D_i$, and this must vanish in A^1X by definition.

Proposition. *For a nonsingular projective toric variety* X,
$A^*X = H^*X = \mathbb{Z}[D_1, \ldots, D_d]/I$, *where* I *is the ideal generated by all*

(i) $D_{i_1} \cdot \ldots \cdot D_{i_k}$ *for* $v_{i_1}, \ldots, v_{i_k}$ *not in a cone of* Δ;

(ii) $\displaystyle\sum_{i=1}^{d} \langle u, v_i \rangle D_i$ *for* u *in* M.

In fact in (i) it suffices to include only the sets of v_i without repeats, and in (ii) one needs only those u from a basis of M.

Proof. Let $A^{\cdot} = \mathbb{Z}[D_1, \ldots, D_d]/I$, with $D_1, \ldots, D_d$ regarded as indeterminates and I generated by the elements in (i) and (ii). By the discussion before the proposition, there is a canonical surjection from the ring $A^{\cdot}$ to A^*X that takes D_i to the class of the corresponding divisor. For each k-dimensional cone σ, let $p(\sigma)$ be the monomial $D_{i_1} \cdot \ldots \cdot D_{i_k}$ in $A^{\cdot}$, where $v_{i_1}, \ldots, v_{i_k}$ are the generators of σ. Choose an ordering of the n-dimensional cones $\sigma_1, \ldots, \sigma_m$ satisfying (∗), with $\tau_1, \ldots, \tau_m$ as before. To complete the proof, i.e., to show that $A^{\cdot} \to A^*X$ is injective, it suffices to show that $p(\tau_1), \ldots, p(\tau_m)$ generate $A^{\cdot}$ as a $\mathbb{Z}$-module. For this we need an "algebraic moving lemma:"

Lemma. *Let* $\alpha \underset{\neq}{\leq} \gamma < \beta$ *be cones in* Δ, *and let* $k = \dim(\gamma)$. *Then there is an equation* $p(\gamma) = \Sigma\, m_i p(\gamma_i)$ *in* $A^{\cdot}$, *with* γ_i *cones of dimension* k *in* Δ *with* $\alpha < \gamma_i$ *but* $\gamma_i \not\subset \beta$, *and* m_i *integers.*

Geometrically, $V(\beta) \subset V(\gamma) \subsetneq V(\alpha)$, and this says we can move $V(\gamma)$ off $V(\beta)$ by a rational equivalence, while staying inside $V(\alpha)$.

Proof. Renumbering $v_1, \ldots, v_d$, we may assume that α is generated by $v_1, \ldots, v_p$, γ by $v_1, \ldots, v_k$, and β by $v_1, \ldots, v_q$, and that $v_1, \ldots, v_n$ form a basis for N, with $1 \leq p < k \leq q \leq n$. Take u in M so that $\langle u, v_k \rangle = 1$ and $\langle u, v_i \rangle = 0$ for $i \leq n$, $i \neq k$. Relation (ii) then gives an equation $D_k = a_{n+1}D_{n+1} + \ldots + a_d D_d$; multiplying by $D_1 \cdot \ldots \cdot D_{k-1}$, we have

$$D_1 \cdot \ldots \cdot D_k = \sum_{i=n+1}^{d} a_i (D_1 \cdot \ldots \cdot D_{k-1}) \cdot D_i \,.$$

Using (i) to throw away any terms for which $v_1, \ldots, v_{k-1}, v_i$ do not generate a cone, this is the required equation.

Next we show that $A^{\cdot}$ is additively generated by monomials in $D_1, \ldots, D_d$ without multiple factors, i.e., by the elements $p(\gamma)$ as γ varies over all cones in Δ. By induction it suffices, if $v_j \in \gamma$, to write $D_j \cdot p(\gamma)$ as a linear combination of such monomials. Use the lemma to write $D_j = \Sigma\, a_i D_i$, a sum over i such that $v_i \notin \gamma$. Then $D_j \cdot p(\gamma) = \Sigma\, a_i D_i \cdot p(\gamma)$ has the required form.

Finally, to see that the $p(\tau_i)$ generate $A^{\cdot}$, by descending induction on i we show that if γ lies between τ_i and σ_i, then $p(\gamma)$ is in the submodule generated by those $p(\tau_j)$ with $j \geq i$. If $\gamma = \tau_i$ we are done, and if not we use the lemma to write $p(\gamma) = \Sigma\, m_t p(\gamma_t)$, a sum over cones γ_t with $\tau_i \subset \gamma_t \not\subset \sigma_i$. By (*), such γ_t must lie between τ_j and σ_j for some $j > i$, and the inductive hypothesis concludes the proof of the proposition.

If Δ is only simplicial, the same holds, with essentially the same proof, but with rational coefficients:

$$\mathbb{Q}[D_1, \ldots, D_d]/I \xrightarrow{\cong} A^*(X)_{\mathbb{Q}} \xrightarrow{\cong} H^*(X; \mathbb{Q}) \,,$$

with I generated by (i) and (ii). The difference in this case is that if τ is generated by $v_{i_1}, \ldots, v_{i_k}$, then $D_{i_1} \cdot \ldots \cdot D_{i_k} = \dfrac{1}{\text{mult}(\tau)} V(\tau)$.

The proposition remains true for arbitrary complete and non-singular (or simplicial) toric varieties. The proof given here works if the maximal cones can be ordered to satisfy the assumption ($*$). Danilov proved it in general by showing that the ring $\mathbb{Z}[D_1, \ldots, D_d]/J$, where J is the ideal generated by the relations (i), is a Gorenstein ring.[12] If X arises from a polytope, this ring is called the *Stanley-Reisner* ring of the polytope.

5.3 Riemann-Roch theorem

We begin by summarizing some general intersection theory. The operation of intersecting with divisors determines, for any line bundle L on any variety X, first Chern class homomorphisms

$$A_k(X) \rightarrow A_{k-1}(X) , \quad \alpha \mapsto c_1(L) \cap \alpha ,$$

defined by the formula $c_1(L) \cap [V] = [E]$, for V a k-dimensional subvariety, where E is a Cartier divisor on V whose line bundle $\mathcal{O}_V(E)$ is isomorphic to the restriction of L to V. From this one can construct, for an arbitrary vector bundle E on X, Chern class homomorphisms $\alpha \mapsto c_i(E) \cap \alpha$ from $A_k(X)$ to $A_{k-i}(X)$, satisfying formulas analogous to those in topology. From these one can define the *Chern character* $\text{ch}(E)$ and the *Todd class* $\text{td}(E)$. These characteristic classes are contravariant for arbitrary maps of varieties. For a line bundle L, these are given by the formulas

$$\text{ch}(L) = \exp(c_1(L)) = 1 + c_1(L) + \ldots + (1/n!)c_1(L)^n ,$$

$n = \dim(X)$; and $\text{td}(L) = c_1(L)/(1 - \exp(-c_1(L))) = 1 + \tfrac{1}{2} c_1(L) + \ldots$. For general bundles, they are determined by requiring them to be multiplicative, i.e., for any short exact sequence $0 \rightarrow E' \rightarrow E \rightarrow E'' \rightarrow 0$ of bundles, $\text{ch}(E) = \text{ch}(E') \cdot \text{ch}(E'')$, and $\text{td}(E) = \text{td}(E') \cdot \text{td}(E'')$. [13]

Every variety X has a "homology Todd class" $\text{Td}(X)$ in $A_*(X)_{\mathbb{Q}}$, (or in Borel-Moore homology $H_*(X;\mathbb{Q})$), which has the form

$$Td(X) = Td_nX + Td_{n-1}X + \ldots + Td_0X ,$$

with $Td_k(X) \in A_k(X)_\mathbb{Q}$. The top class $Td_nX = [X]$ is the fundamental class of X. If X is nonsingular, then

$$Td(X) = td(T_X) \cap [X] ,$$

where $td(T_X)$ is the Todd class of the tangent bundle T_X of X. If $f: X' \to X$ is a proper birational morphism, with $f_*(\mathcal{O}_{X'}) = \mathcal{O}_X$ and $R^i f_*(\mathcal{O}_{X'}) = 0$ for $i > 0$, then $f_*(Td(X')) = Td(X)$. The *Hirzebruch-Riemann-Roch formula* says that for any vector bundle E on a complete variety X,

$$\chi(X,E) = \int ch(E) \cap Td(X) ,$$

where the integral sign means to take the degree of the zero-dimensional piece of the term following. For a line bundle L, it says

$$\chi(X,L) = \sum_{k=0}^{n} \frac{1}{k!} degree(c_1(L)^k \cap Td_k(X)) .$$

For L trivial, this says that $\chi(X,\mathcal{O}_X) = degree(Td_0(X))$. If X is nonsingular, $\chi(X,L)$ is the degree of the top-codimensional piece of $ch(L) \cup td(T_X)$. [14]

When X is a toric variety with divisors $D_1, \ldots, D_d$ as before, the proposition at the end of §4.3 shows that the Chern classes of the cotangent bundle Ω_X^1 and the structure sheaves $\mathcal{O}_{D_i}$ are related by

$$c(\Omega_X^1) \cdot \prod_{i=1}^{d} c(\mathcal{O}_{D_i}) = 1 .$$

Here $c = 1 + c_1 + c_2 + \ldots + c_n$ is the *total Chern class*. From the sequence $0 \to \mathcal{O}(-D_i) \to \mathcal{O}_X \to \mathcal{O}_{D_i} \to 0$, we have $c(\mathcal{O}_{D_i}) = (1 - D_i)^{-1}$. Combining, and using the fact that $c_i(E^\vee) = (-1)^i c_i(E)$, we get

Lemma. *The total Chern class of a nonsingular toric variety is*

$$c(T_X) = \prod_{i=1}^{d} (1 + D_i) = \sum_{\sigma \in \Delta} [V(\sigma)] .$$

By the definition of the Todd class of a bundle, we have similarly

$$td(T_X) \; = \; \prod_{i=1}^{d} \frac{D_i}{1 - \exp(-D_i)}$$

$$= \; 1 + \frac{1}{2} c_1(X) + \frac{1}{12} (c_1(X)^2 + c_2(X)) + \frac{1}{24} c_1(X) \cdot c_2(X) + \dots \;.$$

Here $c_i(X) = c_i(T_X)$. (Note, however, that the D_i are not Chern roots of T_X, since there are more than n of them.) If $X(\Delta)$ is singular, we can compute $Td(X)$ by subdividing to find a proper birational morphism $f: X(\Delta') \to X(\Delta)$ from a nonsingular toric variety. From the last proposition in §3.5 it follows that $Td(X(\Delta)) = f_*(Td(X(\Delta')))$, which gives a method for calculating the Todd class in any given example.

Exercise. For any toric variety X, show that $Td_{n-1}(X) = \frac{1}{2} \sum [D_i]$, and $Td_0(X) = [x]$ for any point x in X. [15]

We will apply the Hirzebruch-Riemann-Roch formula to the line bundle $\mathcal{O}(D)$ of a T-Cartier divisor D on $X(\Delta)$ that is generated by its sections. As we saw in §3.4, the higher cohomology groups of such a line bundle vanish, and the dimension of the space of sections is the number of lattice points in a certain convex polytope $P = P_D$ with vertices in the lattice M. Denote the number of lattice points that are in P by $\#(P)$, i.e., $\#(P) = \text{Card}(P \cap M)$. By Riemann-Roch, we therefore have a formula for the number of such lattice points:

$$\text{(RR)} \qquad \#(P) \; = \; \sum_{k=0}^{n} \frac{1}{k!} \, \text{degree}(D^k \cap Td_k(X)) \;.$$

Note that any bounded convex polytope P with vertices in M arises in this way. Given P, we may find a complete fan Δ that is compatible with P, so there is a T-Cartier divisor D on $X = X(\Delta)$ such that $\mathcal{O}(D)$ is generated by its sections, and so that these sections are precisely the linear combinations of the functions χ^u for u in $P \cap M$. Moreover, by subdividing Δ if necessary, we may assume that $X(\Delta)$ is smooth and projective, although this is rarely necessary.

For any nonnegative integer v, let $v \cdot P = \{v \cdot u : u \in P\}$. From

(RR) and the fact that $v \cdot P$ is the polytope corresponding to the divisor $v \cdot D$ we deduce that $v \mapsto \#(v \cdot P)$ *is a polynomial function of* v of degree at most n:

$$\#(v \cdot P) = \chi(X, \mathcal{O}(v \cdot D)) = \sum_{k=0}^{n} a_k v^k ,$$

$$a_k = \frac{1}{k!} \text{degree}(D^k \cap Td_k(X)) .$$

Let $f_P(v)$ be this polynomial. Note that $f_P(v) = \chi(X, \mathcal{O}(v \cdot D))$ for all $v \in \mathbb{Z}$. Since $Td_n(X) = [X]$, the leading coefficient is $a_n = (D^n)/n!$, where (D^n) denotes the self-intersection number obtained by intersecting D with itself n times.

The lattice M determines a volume element on $M_\mathbb{R}$, by requiring that the volume of the unit cube determined by a basis of M is 1. Denote the volume of a polytope P by $\text{Vol}(P)$. The only fact we will need is the basic and elementary identity that relates the volume to the number of lattice points:

$$(*) \qquad \text{Vol}(P) = \lim_{v \to \infty} \frac{\#(v \cdot P)}{v^n} .$$

This follows from the fact that the error term in estimating the volume by using unit cubes centered at lattice points of a polytope is bounded by the $(n-1)$-dimensional area of the boundary of the polytope, and this vanishes in the limit.

Applying (RR), we see that $\#(v \cdot P)/v^n \to (D^n)/n!$ as $v \to \infty$, where (D^n) is the self-intersection number. Therefore,

Corollary. *If the polytope* P *corresponds to the divisor* D, *then*

$$\text{Vol}(P) = \frac{(D \cdot \ldots \cdot D)}{n!} = \frac{(D^n)}{n!} .$$

The divisor D is described by a collection of elements $u(\sigma)$ in $M/M(\sigma)$, one for each cone σ. For any cone σ in Δ, let P_σ be the intersection of P with the corresponding translation of the subspace perpendicular to σ:

$$P_\sigma = P \cap (\sigma^\perp + u(\sigma)) ;$$

i.e., $P_\sigma = P \cap (\sigma^\perp + u)$ for any u in M that maps to $u(\sigma)$ in $M/M(\sigma)$. The lattice $M(\sigma)$, translated by $u(\sigma)$, determines a volume element on the space $\sigma^\perp + u(\sigma)$, whose dimension is the codimension of σ, or the (complex) dimension of $V(\sigma)$. We have

Corollary. $\mathrm{Vol}(P_\sigma) \;=\; \deg\left(\dfrac{D^k}{k!} \cap [V(\sigma)]\right).$

Proof. Suppose first that $V(\sigma)$ does not belong to the support of D, i.e., $u(\sigma) = 0$. In this case, D restricts to the divisor on the toric variety $V(\sigma)$ defined by the collection of elements $u(\tau) \in M(\sigma)/M(\tau)$ as τ varies over the star of σ. This is a divisor whose line bundle $\mathcal{O}(D|_{V(\sigma)})$ is generated by its sections, and these sections are the linear combinations of characters χ^u for u in $\sigma^\perp \cap P \cap M = P_\sigma \cap M(\sigma)$, so we are reduced to the situation of the preceding corollary. In the general case D can be replaced by $D + \mathrm{div}(\chi^u)$, where u is any element of M that maps to $u(\sigma)$ in $M/M(\sigma)$. We have seen that the polytope corresponding to $D + \mathrm{div}(\chi^u)$ is $P - u$. The first case applies to this, and noting that rationally equivalent divisors have the same intersection numbers, the corollary follows.

In particular, if $X(\Delta_P)$ is constructed from the polytope P as in §1.5, then the cones σ correspond to the faces of P, and in this case P_σ is the face of P corresponding to σ.

We know that the classes of the varieties $V(\sigma)$ span $A_*(X)_{\mathbb{Q}}$, so the Todd class can be written as a $\mathbb{Q}$-linear combination of these classes:

$$Td(X) \;=\; \sum_{\sigma \in \Delta} r_\sigma [V(\sigma)] \,,$$

for some rational numbers r_σ. As we pointed out, in any example one can calculate such coefficients, but because of the relations among these generators for the Chow groups, the coefficients are not unique. Having such an expression for the Todd class, however, gives a marvelous formula for the number of lattice points.[16] Combining (RR) with the preceding corollary, we have

$$\#(P) \;=\; \sum_{\sigma \in \Delta} r_\sigma \cdot \mathrm{Vol}(P_\sigma) \,,$$

and

$$\#(\nu \cdot P) = \sum_{k=0}^{n} a_k \nu^k , \quad \text{with} \quad a_k = \sum_{\text{codim}(\sigma) = k} r_\sigma \cdot \text{Vol}(P_\sigma) .$$

For example, in the two-dimensional case, we may take

$$Td(X) = [X] + \tfrac{1}{2}(\Sigma [D_i]) + [x] ,$$

where x is any point of X. Starting with a convex rational polytope P in the plane, this leads to

Pick's Formula: $\#(P) = \text{Area}(P) + \tfrac{1}{2} \cdot \text{Perimeter}(P) + 1$.

Note that the lengths of the edges of P are measured from the restricted lattice, so the length of an edge between two lattice points is one more than the number of lattice points lying strictly between them.

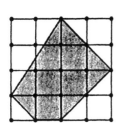

$$\text{Area}(P) = \frac{17}{2}$$

$$\text{Perimeter}(P) = 7$$

$$\#(P) = 13 = \frac{17}{2} + \frac{7}{2} + 1$$

$$\#(\nu \cdot P) = 1 + \frac{7}{2} \nu + \frac{17}{2} \nu^2$$

As we have seen, the value of the polynomial $\nu \mapsto \#(\nu \cdot P)$ at $\nu = -1$ is the number 6 of interior lattice points.

There is great interest in finding useful formulas for the number of lattice points in a convex polytope, which has sparked efforts in finding explicit formulas for the Todd classes of toric varieties. One wants coefficients r_σ that depend only on the geometry around the cone σ. [17] It is also interesting to look for combinatorial formulas for other characteristic classes of a possibly singular toric variety. For example, F. Ehlers has shown that MacPherson's Chern class is always the sum of the classes of all orbit closures, each with coefficient 1, whether the toric variety is singular or not. [18]

Exercise. Let $N = \{(x_0, \ldots, x_n) \in \mathbb{Z}^{n+1} : \Sigma x_i = 0\}$, and define vectors $v_0, \ldots, v_n$ in N whose sum is zero by setting $v_0 = (1, 0, \ldots, -1)$,

$v_1 = (-1,1,0,\ldots,0), \ldots, v_n = (0,\ldots,-1,1)$. Let Δ be the fan with cones generated by proper subsets of $\{v_0,\ldots,v_n\}$, so $X(\Delta) = \mathbb{P}^n$, and each $V(\sigma)$ is an intersection of coordinate hyperplanes in $\mathbb{P}^n$. Show that the coefficient r_σ of $V(\sigma)$ can be taken to be the fraction of σ in the space spanned by σ (by volume obtained by intersecting with the unit ball and using the induced metric from the Euclidean metric on $\mathbb{R}^{n+1}$). [19]

Exercise. Let X be the toric variety whose fan consists of the cones over faces of our standard cube. Compute $\mathrm{Td}(X)$, and use this to compute $\#(v \cdot P)$, where P is the dual octahedron. [20]

Exercise. (R. Morelli) Let $M = \mathbb{Z}^3$, and let P be the polytope with vertices at $(0,0,0)$, $(1,0,0)$, $(0,1,0)$, and $(1,1,m)$, where m is a positive integer. Show that the polynomial $\#(v \cdot P)$ corresponding to P is $1 + \frac{12-m}{6} v + v^2 + \frac{m}{6} v^3$. In particular, if $m \geq 13$, the one-dimensional Todd class $\mathrm{Td}_1(X(\Delta_P))$ cannot be written as a *non-negative* linear combination of the one-dimensional cycles $V(\sigma)$.

5.4   Mixed volumes

Intersection theory on toric varieties can be used to prove interesting facts about convex bodies in Euclidean spaces, by exploiting the interpretation of the volume of a polytope as a self-intersection number of a divisor on a toric variety. The basic notions involved here are Minkowski sums and mixed volumes. If P and Q are any convex compact sets, their *Minkowski sum* is the convex set

$$P + Q = \{u + u' : u \in P, u' \in Q\};$$

note that for a positive integer v, $v \cdot P$ is the Minkowski sum of v copies of P. We work in an n-dimensional Euclidean space $M_{\mathbb{R}}$ coming from a lattice M, which determines a volume $\mathrm{Vol}(P)$ for any compact convex set P. This volume is positive when P has a nonempty interior. For $n = 2$, the mixed volume $V(P,Q)$ of two convex compact sets can be defined by the equation

$$2 \cdot V(P,Q) = \mathrm{Vol}(P + Q) - \mathrm{Vol}(P) - \mathrm{Vol}(Q),$$

where of course Vol denotes ordinary area. In particular, V(P,P) is
the area of P. For example, if Q = B(ε) is the disk of radius ε, then
P + Q is a closed ε-neighborhood of P, and the right side of this
equation, when divided by ε, approaches the circumference of P as
ε approaches 0.

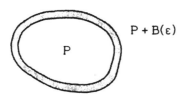

The mixed volume is multilinear, so V(P,B(ε)) = ε·V(P,B), where B
is the unit disk. It follows that 2·V(P,B) is the circumference of P.
The isoperimetric inequality that the area of P is at most the square
of the circumference divided by 4π translates to the inequality
V(P,Q)^2 ≥ V(P,P)·V(Q,Q).

Our goal in this section is to prove generalizations of these facts.
To start, we will consider only n-dimensional convex polytopes whose
vertices are in M. All of the assertions, in fact, will extend to general
convex compact sets, by approximating them with such rational
polytopes (for a finer lattice). We have seen that if a complete fan Δ
is compatible with a polytope P, i.e., the corresponding function ψ_P is
linear on the cones of Δ, then P corresponds to a T-Cartier divisor
D on the toric variety X = X(Δ). The line bundle Ø(D) is generated
by its sections, which have a basis of characters χ^u for u in P∩M.
As we saw at the end of §3.4, the assumption that P is n-dimensional
implies that the map from X to projective space given by these
sections has an image that is an n-dimensional variety.

If any finite number of such polyhedra $P_1, \ldots, P_s$ are given, we
may find one such Δ that is compatible with each of them; if desired,
by subdividing the fan, we may even make X = X(Δ) nonsingular and
projective. If E_i is the divisor on X corresponding to the polytope
P_i, then, for any nonnegative integers $v_1, \ldots, v_s$, the polytope
$v_1 \cdot P_1 + \ldots + v_s \cdot P_s$ corresponds to the divisor $v_1 \cdot E_1 + \ldots + v_s \cdot E_s$.
From the corollary in the preceding section we deduce a fundamental
result of Minkowski that the volume of a nonnegative linear

combination of polytopes is a polynomial of degree n in the coefficients $v_1, \ldots, v_s$. In fact, we have the formula

(1) $\mathrm{Vol}(v_1 \cdot P_1 + \ldots + v_s \cdot P_s) = \dfrac{(v_1 \cdot E_1 + \ldots + v_s \cdot E_s)^n}{n!}$.

For n polytopes $P_1, \ldots, P_n$, $n!$ times the coefficient of $v_1 \cdot \ldots \cdot v_n$ in this polynomial is defined to be the *mixed volume* of $P_1, \ldots, P_n$, and is denoted $V(P_1, \ldots, P_n)$. We will give a closed formula for this in a minute, but what will be useful is this intersection-theoretic characterization:

(2) $V(P_1, \ldots, P_n) = \dfrac{(E_1 \cdot \ldots \cdot E_n)}{n!}$.

The mixed volume has a simple expression as an alternating sum of ordinary volumes:

(3) $n! \cdot V(P_1, \ldots, P_n) = \mathrm{Vol}(P_1 + \ldots + P_n) - \sum_{i=1}^{n} \mathrm{Vol}(P_1 + \ldots + \hat{P}_i + \ldots + P_n)$

$\qquad + \sum_{i < j} \mathrm{Vol}(P_1 + \ldots + \hat{P}_i + \ldots + \hat{P}_j + \ldots + P_n) - \ldots + (-1)^{n-1} \sum_{i=1}^{n} \mathrm{Vol}(P_i)$.

This follows from (2) and the algebraic identity

$\qquad n! \cdot (E_1 \cdot \ldots \cdot E_n) = (E_1 + \ldots + E_n)^n - \sum_{i=1}^{n} (E_1 + \ldots + \hat{E}_i + \ldots + E_n)^n$

$\qquad + \sum_{i < j} (E_1 + \ldots + \hat{E}_i + \ldots + \hat{E}_j + \ldots + E_n)^n - \ldots + (-1)^{n-1} \sum_{i=1}^{n} (E_i)^n$.

(Here and in the following, to avoid a flood of parentheses, we often write E^n in place of (E^n) for a divisor E; it is always the intersection number that is meant.) Many other properties of mixed volume are easy to prove either from (2), or directly from the definition. For example, we see immediately that $V(P_1, \ldots, P_n)$ is a symmetric and multilinear function of its variables:

(4) $V(a \cdot P + b \cdot Q, P_2, \ldots, P_n) = a \cdot V(P, P_2, \ldots, P_n) + b \cdot V(Q, P_2, \ldots, P_n)$

for nonnegative integers a and b. It also follows from this definition
that

(5) $V(P, \ldots, P) = \text{Vol}(P)$.

Exercise. Prove, for two polytopes P and Q, the *Steiner
decomposition:*

(6) $\text{Vol}(P + Q) = \sum_{i=0}^{n} \binom{n}{i} V(P,i; Q, n-i)$,

where $V(P,i; Q, n-i)$ denotes the mixed volume of i copies of P and
n-i copies of Q.

The mixed volume is invariant under arbitrary translations of
the polytopes:

(7) $V(P_1 + u_1, \ldots, P_n + u_n) = V(P_1, \ldots, P_n)$

for any $u_1, \ldots, u_n$ in M. This follows from the fact that $P_i + u_i$
corresponds to the divisor $E_i - \text{div}(\chi^{u_i})$, because intersection numbers
of rationally equivalent divisors are the same.

Exercise. If M is replaced by $\frac{1}{m} M$, show that the mixed volumes
are all multiplied by m^n. Show that

(8) $V(A(P_1), \ldots, A(P_n)) = |\det(A)| \cdot V(P_1, \ldots, P_n)$

for any linear transformation A of M.

An important fact that is far less obvious is the *nonnegativity* of
mixed volumes. This follows from the fact that the intersection
number of n divisors E_i, when each $\mathcal{O}(E_i)$ is generated by its
sections, is nonnegative. The following is a stronger result:

(9) $V(Q_1, \ldots, Q_n) \leq V(P_1, \ldots, P_n)$ if $Q_i \subset P_i$ for $1 \leq i \leq n$.

It suffices to prove this when $Q_i = P_i$ for $i \geq 2$. If E_i are the divisors
corresponding to P_i, and F_1 corresponds to Q_1, the fact that Q_1 is
contained in P_1 implies that $E_1 = F_1 + Y$ with Y an effective
divisor. Changing E_i to rationally equivalent divisors if necessary, we
may assume each E_i restricts to a Cartier divisor on Y. Then

$(E_1 \cdot E_2 \cdot \ldots \cdot E_n) - (F_1 \cdot E_2 \cdot \ldots \cdot E_n) = (Y \cdot E_2, \ldots \cdot E_n) = (E_2|_Y \cdot \ldots \cdot E_n|_Y)$.

This is nonnegative since it is the intersection number of n-1 divisors on Y whose line bundles $\mathcal{O}(E_i|_Y)$ are generated by their sections.[21] In particular, if P is the smallest convex set containing $P_1, \ldots, P_n$, (9) gives the inequality

(10) $V(P_1, \ldots, P_n) \leq \text{Vol}(P)$.

We also derive a stronger version of nonnegativity:

(11) $V(P_1, \ldots, P_n) > 0$ if $\text{Int}(P_i) \neq \emptyset$ for $1 \leq i \leq n$.

To see this, replace M by $\frac{1}{m}$ M and translate the polytopes if necessary so that the intersection $Q = \bigcap P_i$ has nonempty interior, and then (9) and (5) imply that the mixed volume is at least the volume of Q.

For a simple example, let P_1 and P_2 be the polyhedra in $\mathbb{R}^2$ with vertices in $\mathbb{Z}^2$ as shown:

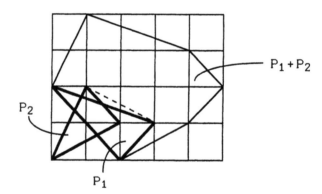

Then $\text{Vol}(P_1) = 2$, $\text{Vol}(P_2) = 1\frac{1}{2}$, and $\text{Vol}(P_1+P_2) = 11\frac{1}{2}$, so

$$V(P_1, P_2) = \frac{1}{2}(\text{Vol}(P_1+P_2) - \text{Vol}(P_1) - \text{Vol}(P_2)) = 4 .$$

The volume of the convex hull P is $4\frac{1}{2}$.

Exercise. Prove the *additivity formula:* given P and Q such that $P \cup Q$ is convex, and $P_{k+1}, \ldots, P_n$, then

(12) $V(P \cup Q, k; P_{k+1}, \ldots, P_n) + V(P \cap Q, k; P_{k+1}, \ldots, P_n)$

$$= V(P, k; P_{k+1}, \ldots, P_n) + V(Q, k; P_{k+1}, \ldots, P_n) ,$$

where the notation means that the first polytope is repeated k times.

Finally, we have a deeper result, known as the *Alexandrov-Fenchel inequality:*

(13) $V(P_1, \ldots, P_n)^2 \geq V(P_1, P_1, P_3, \ldots, P_n) \cdot V(P_2, P_2, P_3, \ldots, P_n)$.

Following Teissier and Khovanskii, this can be deduced from the Hodge index theorem. If $E_1, \ldots, E_n$ are the corresponding divisors on the variety X, which we can take to be nonsingular, (13) is equivalent to the inequality

$$(E_1 \cdot \ldots \cdot E_n)^2 \geq (E_1 \cdot E_1 \cdot E_3 \cdot \ldots \cdot E_n)(E_2 \cdot E_2 \cdot E_3 \cdot \ldots \cdot E_n) .$$

The maps $\varphi_i \colon X \to \mathbb{P}^{r_i}$ determined by the divisors E_i have n-dimensional images. The Bertini theorem[22] implies that if H_i is a generic hyperplane in $\mathbb{P}^{r_i}$, then $Y = \varphi_3^{-1}(H_3) \cap \ldots \cap \varphi_n^{-1}(H_n)$ is a nonsingular irreducible surface. Since $\varphi_i^{-1}(H_i)$ is rationally equivalent to E_i, if we define D_1 and D_2 to be the restrictions of E_1 and E_2 to the surface Y, this inequality becomes

$$(D_1 \cdot D_2)^2 \geq (D_1 \cdot D_1)(D_2 \cdot D_2) .$$

This is a consequence of the Hodge index theorem, which says that if A and B are divisors on a nonsingular projective surface, with $(A \cdot A) > 0$ and $(A \cdot B) = 0$, then $(B \cdot B) \leq 0$.[23] To deduce the displayed inequality, we know that $(D_1 \cdot D_1) \geq 0$, and the inequality is obvious if $(D_1 \cdot D_1) = 0$. If $(D_1 \cdot D_1) > 0$, take $A = D_1$ and $B = aD_2 - bD_1$, with $a = (D_1 \cdot D_1)$ and $b = (D_1 \cdot D_2)$; the index theorem gives the inequality $(aD_2 - bD_1)^2 \leq 0$, which is immediately seen to be equivalent to the displayed equation. Note that if $n = 2$, the Bertini argument is unnecessary.

Exercise. (a) Given $P, Q, R_{k+1}, \ldots, R_n$, and $0 < i < k$, prove the inequality

$$V(P, i; Q, k-i; R_{k+1}, \ldots, R_n)^k \geq V(P, k; R_{k+1}, \ldots, R_n)^i \cdot V(Q, k; R_{k+1}, \ldots, R_n)^{k-i} .$$

(b) Show that $V(P_1, \ldots, P_n)^k \geq \prod_{i=1}^{k} V(P_i, k; P_{k+1}, \ldots, P_n)$.

(c) Deduce the inequality

(14) $V(P_1, \ldots, P_n)^n \geq \mathrm{Vol}(P_1) \cdot \ldots \cdot \mathrm{Vol}(P_n)$.[24]

The inequality (14) is equivalent to the inequality

$$(E_1 \cdot \ldots \cdot E_n)^n \geq (E_1{}^n) \cdot \ldots \cdot (E_n{}^n) .$$

This is also a general fact, for divisors E_i such that each $\mathcal{O}(E_i)$ is generated by its sections, on any complete irreducible n-dimensional variety. This appealing generalization of the Hodge index theorem was apparently only recently discovered by Demailly.[25]

Exercise. Prove the *Brunn-Minkowski inequality* for two convex polytopes P and Q:

$$\sqrt[n]{\mathrm{Vol}(P+Q)} \geq \sqrt[n]{\mathrm{Vol}(P)} + \sqrt[n]{\mathrm{Vol}(Q)} . \text{(26)}$$

Now let $M = \mathbb{Z}^n$, so $M_\mathbb{R} = \mathbb{R}^n$, with its usual metric. Let B denote the ball of radius one, and $B(\varepsilon)$ the ball of radius ε. All of the preceding results extend to arbitrary compact convex sets, by the following exercise.

Exercise. Define a distance $d(P,Q)$ between compact convex sets to be the minimum ε such that $P \subset Q + B(\varepsilon)$ and $Q \subset P + B(\varepsilon)$. Show that d defines a metric (with triangle inequality). Show that the operations $P, Q \mapsto P + Q$; $P \mapsto \mathrm{Vol}(P)$; $P_1, \ldots, P_n \mapsto V(P_1, \ldots, P_n)$ are continuous. Show that for any compact convex Q and any positive ε there is a positive integer m and an n-dimensional convex polytope P with vertices in $\frac{1}{m} M$ such that $d(P,Q) \leq \varepsilon$.

The mixed volumes are particularly interesting when taken with several copies of one compact convex set and the other copies a ball. For example, the expansion

$$\mathrm{Vol}(P + B(\varepsilon)) = \sum_{i=0}^{n} \binom{n}{i} V(P,i;B(\varepsilon),n-i) = \sum_{i=0}^{n} \binom{n}{i} V(P,i;B,n-i)\varepsilon^{n-i}$$

interprets the mixed volumes of P and the unit ball B in terms of the rate of growth of the volume of an ε-neighborhood of P:

$$\frac{d^k}{d^k t} \mathrm{Vol}(P + B(t))\Big|_{t=0} = n(n-1) \cdot \ldots \cdot (n-k+1) \cdot V(P,n-k;B,k) .$$

In particular, the first derivative of this function, which we denote by $S(P)$, is given by the formula

$$S(P) = n \cdot V(P, \ldots, P, B) .$$

So $S(P)$ is the limit of $(\text{Vol}(P + B(\varepsilon)) - \text{Vol}(P))/\varepsilon$ as ε goes to 0. When the boundary of P is smooth, $S(P)$ is its $(n-1)$-dimensional area; if not, this is a reasonable measure for it. Inequality (14) then implies a classical *isoperimetric inequality*

$$S(P)^n \geq n^n \cdot v_n \cdot \text{Vol}(P)^{n-1} ,$$

where v_n is the volume of B.

It is also interesting to interchange the roles of P and B, since the limit of $(\text{Vol}(B + \varepsilon \cdot P) - \text{Vol}(B))/\varepsilon$ as ε goes to 0 can be written as $S(B) \cdot d(P)/2$, where $d(P)$ is the *mean diameter* of P. In other words, $d(P) = 2 \cdot n \cdot V(P, B, \ldots, B)/S(B)$, and (14) gives the inequality

$$d(P)^n \geq 2^n \cdot n^n \cdot \text{Vol}(P) \cdot \text{Vol}(B)^{n-1}/S(B)^n = 2^n \cdot \text{Vol}(P)/\text{Vol}(B) .$$

In particular, this gives the inequality $\text{Vol}(P) \leq (d/2)^n \cdot \text{Vol}(B)$, where d is the maximal diameter of P.

Exercise. Show that $f(t) = V(t \cdot P + (1-t)Q, k; R_{k+1}, \ldots, R_n)^{1/k}$ has the upper convexity property: $f(t) \geq (1-t)f(0) + t f(1)$. In particular,

$$S(P + Q)^{1/(n-1)} \geq S(P)^{1/(n-1)} + S(Q)^{1/(n-1)} . \quad (27)$$

5.5 Bézout theorem

A regular function on the torus $T = \text{Hom}(M, \mathbb{C}^*)$ has the form $F = \Sigma a_u \chi^u$, the sum over a finite number of points u in M. With $M = \mathbb{Z}^n$, this is a *Laurent polynomial*, i.e., a finite linear combination of monomials in variables $X_1, \ldots, X_n$ with arbitrary integer exponents. The convex hull of the points u for which a_u is not zero is called the *Newton polytope* or *polyhedron* of F. [28] We need the following simple fact.

Lemma. *Let Δ be a complete fan compatible with the Newton*

polytope P *of* F, *and let* E *be the T-Cartier divisor on* $X(\Delta)$
corresponding to P. *Then* $E + \text{div}(F)$ *is an effective divisor on* $X(\Delta)$.

Proof. The assertion is that the order of $E + \text{div}(F)$ along any
codimension one subvariety D of $X(\Delta)$ is nonnegative. This is clear
if D meets T, since F is regular on T. Otherwise D is the
subvariety corresponding to an edge of Δ. With v the generator of
this edge, we have

$$\text{ord}_D(F) \geq \min_{a_u \neq 0} \text{ord}_D(\chi^u) = \min_{u \in P \cap M} \langle u, v \rangle = \psi_P(v) = -\text{ord}_D(E) ,$$

as required.

Suppose $F_1, \ldots, F_n$ are Laurent polynomials. We want to
estimate the number of their common zeros in $T \cong (\mathbb{C}^*)^n$. Let

$$Z = \{z \in T : F_1(z) = \ldots = F_n(z) = 0\} .$$

If z is an isolated point of Z, let $i(z; F_1, \ldots, F_n)$ denote the intersection multiplicity of the hypersurfaces defined by the F_i at the
point z; this intersection number is a positive integer, and it is 1
exactly when the hypersurfaces meet transversally at z. Our Bézout
formula, generalizing slightly results of D. N. Bernstein, A. G.
Kouchnirenko, and A. G. Khovanskii,[29] is

Proposition. *Let* P_i *be the Newton polytope of* F_i. *Then*

$$\sum_{z \text{ isolated in } Z} i(z; F_1, \ldots, F_n) \leq n! \cdot V(P_1, \ldots, P_n) ,$$

where $V(P_1, \ldots, P_n)$ *is the mixed volume.*

Proof. Take Δ compatible with all of the polytopes, so each P_i
corresponds to a T-Cartier divisor E_i on $X(\Delta)$. We have seen that
the right side of the inequality is the intersection number $(E_1 \cdot \ldots \cdot E_n)$.
Since $D_i = \text{div}(F_i) + E_i$ is rationally equivalent to E_i, this intersection
number is the same as $(D_1 \cdot \ldots \cdot D_n)$. Each divisor D_i meets the torus
T in the hypersurface defined there by $F_i = 0$, so the left side of the
inequality is at most the sum of the intersection multiplicities of the
divisors $D_1, \ldots, D_n$ at their isolated points of intersection. The line

bundles $\mathcal{O}(D_i) \cong \mathcal{O}(E_i)$ are generated by their sections, so we are reduced to a general fact of intersection theory: if $D_1, \ldots, D_n$ are effective Cartier divisors on a complete irreducible variety, whose line bundles $\mathcal{O}(D_i)$ are generated by their sections, then the sum of their intersection multiplicities at their isolated points of intersection is at most the intersection number $(D_1 \cdot \ldots \cdot D_n)$. [30]

Note that it is not necessary to assume that Z is finite; and even if Z is finite, the divisors D_i, which represent the closures of the hypersurfaces $F_i = 0$, can have non-isolated intersections on the complement of the torus.

If the polytopes P_i are fixed, it is also true that, for generic F_i with these P_i as Newton polygons, the intersections are all isolated and transversal, and equality holds in the preceding display. This follows from the fact that, in this case, the divisors D_i will be the inverse images of generic hyperplanes for the corresponding maps from $X(\Delta)$ to projective spaces, and the Bertini theorem guarantees that these will have only isolated transversal intersections.

For an example with $n = 2$, consider the polynomials

$$F_1 = X^2 + Y^2 + X^3Y \quad \text{and} \quad F_2 = X^2Y + XY^2 + 1 .$$

The Newton polygons P_1 and P_2 are those sketched in the preceding section. Since the mixed volume $V(P_1, P_2)$ is 4, the number of common isolated zeros must be at most 8. Early estimates bounded the number of zeros by $n! \cdot \mathrm{Vol}(P)$, where P is the convex hull of $P_1, \ldots, P_n$. As we saw in the preceding section, the mixed volume gives a sharper estimate. In the above example, the volume of this convex hull is 4½, so that estimate would allow 9 solutions.

Exercise. Show that there are exactly 8 common solutions to these equations in $(\mathbb{C}^*)^2$, so all of the intersection numbers must be 1.

Consider the case of n polynomials F_i in $\mathbb{C}[X_1, \ldots, X_n]$, with $\deg(F_i) \le d_i$. The Newton polytope of F_i is contained in $d_i \cdot P$, where $P = \{(t_1, \ldots, t_n) \in \mathbb{R}^n : t_i \ge 0 \text{ and } \Sigma t_i \le 1\}$. The standard fan for $\mathbb{P}^n$, regarded as the compactification for $\mathbb{C}^n$ as usual, is compatible with these polytopes, so the sum of the intersection numbers at isolated zeros in $(\mathbb{C}^*)^n$ — or $\mathbb{C}^n$ or $\mathbb{P}^n$ — is at most

$$n! \cdot V(d_1 \cdot P, \ldots, d_n \cdot P) \; = \; n! \cdot d_1 \cdot \ldots \cdot d_n \cdot V(P, \ldots, P)$$
$$= \; n! \cdot d_1 \cdot \ldots \cdot d_n \cdot \text{Vol}(P) \; = \; d_1 \cdot \ldots \cdot d_n \,,$$

since $\text{Vol}(P) = 1/n!$, so the usual Bézout theorem is recovered.

5.6 Stanley's theorem

Stanley has given a beautiful application of toric varieties to the problem of characterizing the numbers of vertices, edges, and faces of all dimensions of a convex simplicial polytope K in Euclidean n-space.

Let K be a convex polytope in 3-space, with f_0 vertices, f_1 edges, and f_2 faces. To begin, we have *Euler's formula:*

(1) $f_0 - f_1 + f_2 \; = \; 2$.

If the polytope is simplicial, i.e., all of its faces are triangles, the facts that each face has three edges and each edge is on two faces give

(2) $3f_2 \; = \; 2f_1$.

To bound a solid takes more than three vertices, i.e.,

(3) $f_0 \; \geq \; 4$.

Note that the triple is determined once the number of vertices is specified. It is an easy exercise to show that any triple of integers (f_0, f_1, f_2) satisfying these three conditions can be realized from a convex simplicial polytope in 3-space. Starting with a solid simplex (with $f_0 = 4$), it suffices to show how to construct a new polytope with one more vertex than a given one. This is achieved by putting a new vertex just outside the middle of a face, and taking the convex hull of the new set of vertices.

For $n = 4$, the numbers f_0, f_1, f_2, f_3 satisfy Euler's equation:

(1) $f_0 - f_1 + f_2 - f_3 \; = \; 0$.

Since 3-simplices have four 2-faces, each on two 3-simplices,

(2) $f_2 \; = \; 2f_3$.

To bound a 4-dimensional solid, as before,

(3) $f_0 \geq 5$.

There is also a quadratic inequality, valid in all dimensions, which comes from the fact that two vertices can be joined by at most one edge:

(4) $f_1 \leq \frac{1}{2} f_0 (f_0 - 1)$.

This time there is also a less obvious lower bound on the number of edges:

(5) $f_1 \geq 4 f_0 - 10$.

For example, if $f_0 = 5$, these conditions determine the other numbers: $f_1 = 10$, $f_2 = 10$, and $f_3 = 5$. This example is achieved by (the boundary of) a 4-simplex.

Exercise. The conditions (1)-(5) allow two possibilities for (f_0, f_1, f_2, f_3) with six vertices: $(6,14,16,8)$ and $(6,15,18,9)$. Construct simplicial polytopes to realize these two possibilities.

The claim is that these five conditions are necessary and sufficient for the existence of a simplicial 4-polytope. In general, the problem is to characterize the n-tuples $(f_0, f_1, \ldots, f_{n-1})$ of integers that can be realized as the numbers of vertices, edges, . . . , $(n-1)$-faces of an n-dimensional convex simplicial polytope. Although it is easy to generalize some of the equations listed above for polytopes of dimension three and four — at least for equations (1)-(4) — it is far from obvious what the complete answer should be, or how to prove it.

There is a convenient way to rewrite the equations, which suggests generalizations, by looking at successive differences. This replaces the sequence $(f_0, f_1, \ldots, f_{n-1})$ by an equivalent sequence of integers $(h_0, h_1, \ldots, h_n)$. Write the integers f_i down the right side of a triangle, and the integer 1 down the left, and then fill in the spaces from the top down so that each integer inside is the difference of the integer above it to the right and that above and to the left:

$$1 \qquad f_0$$
$$1 \qquad f_0-1 \qquad f_1$$
$$1 \qquad f_0-2 \qquad f_1-f_0+1 \qquad f_2$$
$$1 \qquad f_0-3 \qquad f_1-2f_0+3 \qquad f_2-f_1+f_0-1$$

Denote the bottom row, from right to left, by $(h_0, h_1, \ldots, h_n)$. In formulas, the relations are simply

$(*) \qquad h_p = \sum_{i=p}^{n} (-1)^{i-p} \binom{i}{p} f_{n-i-1}$,

where we set $f_{-1} = 1$.

Euler's equation is equivalent to the equation

$$h_0 = h_n .$$

For $n = 3$ and $n = 4$, equations (1) and (2) are equivalent to the equations

$(A) \qquad h_0 = h_n \quad$ and $\quad h_1 = h_{n-1}$.

The generalization $f_0 \geq n+1$ of equation (3) is equivalent to $h_{n-1} \geq 1$, or, modulo the above, to

$(B_1) \qquad h_1 \geq h_0 = 1$.

Equation (5) for $n = 4$ is equivalent, modulo the equations (A), to

$(B_2) \qquad h_2 \geq h_1$,

while equation (4) is equivalent to

$(C) \qquad h_2 - h_1 \leq \binom{h_1 - h_0 + 1}{2}$.

The *Dehn-Sommerville* equations are the generalizations of the equations (A): $h_p = h_{n-p}$. They were expressed in this form by Sommerville in the 1920's. The generalizations of the equations (B) are not hard to guess, but (C) is more subtle. The complete answer is given in the following theorem, which was conjectured by P. McMullen.

Given a sequence $f_0, f_1, \ldots, f_{n-1}$ of integers, set $f_{-1} = 1$ and $m = [\frac{n}{2}]$. For $0 \le p \le n$, define h_p by (*), and, for $1 \le p \le m$, set

$$g_p = h_p - h_{p-1} .$$

Theorem. *A sequence of integers* $f_0, f_1, \ldots, f_{n-1}$ *occurs as the number of vertices, edges, ... , (n-1)-faces of an n-dimensional convex simplicial polytope if and only if the following conditions hold:*

(I) *(Dehn-Sommerville)* $h_p = h_{n-p}$ *for* $0 \le p \le m$;

(II) $(g_1, \ldots, g_m)$ *is a Macaulay vector, i.e.,*

(a) $g_p \ge 0$ *for* $1 \le p \le m$,

and, if one writes

$$g_p = \binom{n_p}{p} + \binom{n_{p-1}}{p-1} + \ldots + \binom{n_r}{r} ,$$

with $n_p > n_{p-1} > \ldots > n_r \ge r \ge 1$, *then*

(b) $g_{p+1} \le \binom{n_p+1}{p+1} + \binom{n_{p-1}+1}{p} + \ldots + \binom{n_r+1}{r+1} ,$

for $1 \le p \le m - 1$.

Note that for any positive integer p, any positive integer g_p has a unique expression in the form (a), by taking n_p maximal such that $\binom{n_p}{p} \le g_p$, and then continuing with the difference. In words, (b) says that the bound for g_{p+1} is obtained from the expression for g_p by increasing each *entry* in each binomial number by one. If $g_p = 0$, the condition is interpreted to mean that $g_{p+1} = 0$. The existence of a convex polytope with such face numbers was established by Billera and Lee by a direct ingenious argument; the necessity was proved by Stanley.[31] We give Stanley's argument.

The main idea is to produce a toric variety $X = X(\Delta)$ for some complete simplicial fan Δ, so that the number d_k of k-dimensional cones in Δ is the same as the number f_{k-1} of (k-1)-dimensional faces of the given simplicial polytope. To do this, note that, since the polytope is simplicial, moving all of its vertices slightly gives a polytope

with the same numbers of faces in each dimension. By such a
perturbation we can assume the vertices are in the rational points N_Q
of a given lattice N, or by refining N, that the vertices are all in N.
We may also translate so that the origin is in the interior of the
polytope. Now define Δ to be the set of cones over the faces of the
polytope (with vertex at the origin), together with the cone (0). The
equation $d_k = f_{k-1}$ (with $d_0 = f_{-1} = 1$) is then clear. The variety
$X = X(\Delta)$ is complete, and locally a quotient of $\mathbb{C}^n$ by a finite abelian
group, since Δ is simplicial. In addition, X is projective. In fact, if
P in M_R is the polar polytope to the given polytope, then $\Delta = \Delta_P$,
and $X = X(\Delta_P)$ comes equipped with an ample line bundle.

 We can now apply the theorems about the cohomology of the
toric variety X from the first part of this chapter. The rational
cohomology $H^*(X) = H^*(X;\mathbb{Q})$ vanishes in odd dimensions, and

$$h_p = \dim H^{2p}(X) = \sum_{i=p}^{n} (-1)^{i-p} \binom{i}{p} d_{n-i} = \sum_{i=p}^{n} (-1)^{i-p} \binom{i}{p} f_{n-i-1} .$$

We saw that (I) is a corollary of Poincaré duality, and that (II)(a)
follows from the hard Lefschetz theorem for intersection homology. To
prove (II)(b), we use the other general fact we know about the
cohomology of X: it is generated by classes of divisors. Set

$$R^i = H^{2i}(X;\mathbb{Q})/\omega \cdot H^{2i-2}(X;\mathbb{Q}) ,$$

where, as before, ω is the class of a hyperplane. Then $R^* = \oplus R^i =$
$H^*(X)/(\omega)$ is a commutative graded $\mathbb{Q}$-algebra, with $R^0 = \mathbb{Q}$, that is
generated by the part R^1 of degree one. Macaulay characterized the
sequences $(g_1, g_2, \ldots)$ of integers that can be the dimensions of such
an algebra: they must form a Macaulay vector as in (II). [32] Noting
that

$$g_p = \dim R^p = \dim H^{2p}(X) - \dim H^{2p-2}(X) = h_p - h_{p-1}$$

for $1 \le p \le m$, the proof of the theorem is complete.

Exercise. (a) Show that the relation between the f_p's and the h_p's
can be expressed by the polynomial identity

$$\sum_{p=0}^{n} h_p t^p = \sum_{i=0}^{n} f_{n-i-1}(t-1)^i .$$

(b) Show that

$$f_{p-1} = \sum_{q=0}^{p} \binom{n-q}{n-p} h_{n-q} \quad \text{for } 0 \le p \le n .$$

(c) Show that the Dehn-Sommerville equations (I) are equivalent to the equations:

$$f_p = (-1)^{n-1} \sum_{k=p}^{n-1} (-1)^k \binom{k+1}{p+1} f_k \quad \text{for } -1 \le p < n .$$

Exercise. For $n = 5$, use (I) to solve for f_2, f_3, and f_4 in terms of f_0 and f_1, and write (II) as inequalities for f_0 and f_1.

Exercise. Prove that

$$h_p \le \binom{h_1 + p + 1}{p} = \binom{v - n + p - 1}{p} \quad \text{for } 0 \le p \le n ,$$

where $v = f_0$ is the number of vertices. (Here the convention is that $\binom{a}{b}$ takes its usual values for $0 \le b \le a$, and is otherwise 0, except that $\binom{-1}{0} = 1$.) Deduce the Motzkin-McMullen *upper bound conjecture*:

$$f_p \le \sum_{k=0}^{m} \frac{v}{v-k} \binom{v-k}{k} \binom{k}{p+1-k} \quad \text{when } n = 2m ;$$

$$f_p \le \sum_{k=0}^{m} \frac{p+2}{v-k} \binom{v-k}{k+1} \binom{k+1}{p+1-k} \quad \text{when } n = 2m+1 .$$

Prove that these inequalities are valid for any convex n-dimensional polytope, simplicial or not. Show that these inequalities become equalities for a (simplicial) *cyclic* polytope: the convex hull of v points on the rational normal curve $\{(t, t^2, t^3, \ldots, t^n)\}$. [33]

Exercise. Deduce the *lower bound conjecture:* for a simplicial polytope,

$$f_p \ge \binom{n}{p} f_0 - \binom{n+1}{p+1} p \quad \text{for } 1 \le p \le n-2 ;$$

$$f_{n-1} \ge (n-1) f_0 - (n+1)(n-2) .$$

Exercise. A convex n-dimensional polytope is called *simple* if each vertex lies on exactly n (n-1)-faces. Characterize the sequences $(f_0, \ldots, f_{n-1})$ of integers that arise from a simple polytope.[34]

These links between toric varieties and polytopes have spurred renewed interest in both subjects.[35]

NOTES

Preface

1. See for example,

   E. Arbarello, M. Cornalba, P. A. Griffiths, and J. Harris, *Geometry of Algebraic Curves, Vol. I,* Springer-Verlag, 1985.

2. See [Beau], and

   W. Barth, C. Peters, and A. Van de Ven, *Compact Complex Surfaces,* Springer-Verlag, 1984.

3. See

   J. Harris, *Algebraic Geometry: A First Course,* Springer-Verlag, 1992.

4. The Danilov article is [Dani]. The others are [Bryl], [Jurk], [Dema], [KKMS], [AMRT], and [Oda'].

5. Again Oda has come to the rescue, with a survey:

   T. Oda, "Geometry of toric varieties," pp. 407-440, in *Proc. of the Hyderabad Conf. on Algebraic Groups,* 1989 (S. Ramanan, ed.), Manoj Prakashan, Madras, 1991.

A new text emphasizes convex geometry:

   G. Ewald, *Combinatorial Geometry and Algebraic Geometry,* to appear.

Chapter 1

1. See [Dema], [KKMS], and

   I. Satake, "On the arithmetic of tube domains," Bull. Amer. Math. Soc. **79** (1973), 1076-1094.

   Y. Namikawa, *Toroidal Compactification of Siegel Spaces,* Springer Lecture Notes **812**, 1980.

2. The first assertion is seen by writing down the transition functions for the bundle. See [Shaf, Ch. VI] for the basic facts about vector bundles, including $\mathcal{O}(a)$.

3. Let $\mathcal{I}$ be the ideal sheaf of D_τ in $\mathbb{F}_a$. On U_{σ_1} the conormal sheaf

$\mathcal{I}/\mathcal{I}^2$ is trivial and generated by Y; and on U_{σ_4} it is trivial and generated by X^aY, so the change of coordinates is by X^a, which identifies $\mathcal{I}/\mathcal{I}^2$ with $\mathcal{O}(a)$. The self-intersection number is the degree of the normal bundle $\mathcal{O}(-a)$. The self-intersection numbers for the other three curves are 0, a, and 0. We will describe the divisors corresponding to rays more fully in Chapter 3, and intersection numbers in Chapter 5. See [Beau] for basic facts about the surfaces $\mathbb{F}_a$.

4. Some of these references are:

> A. Brøndsted, *An Introduction to Convex Polytopes,*
>   Springer-Verlag, 1983.

> B. Grünbaum, *Convex Polytopes,* J. Wiley and Sons, 1967.

> R. T. Rockafellar, *Convex Analysis,* Princeton Univ. Press, 1970.

The Appendix in [Oda] has a more complete list of facts relevant to toric varieties, with precise references.

5. See for example Lemma 4.4 in the book by Brøndsted, §2.2 in that by Grünbaum, or Theorem 11.1 in that by Rockafellar.

6. See [Oda, Lemma A.4], and the article

> R. Swan, "Gubeladze's proof of Anderson's conjecture," pp. 215-250
>   in *Azumaya Algebras, Actions, and Modules,* Contemp.
>   Math. **124**, Amer. Math. Soc., 1992.

7. Not in dimensions larger than three. For example, if σ is the cone in $\mathbb{Z}^4$ generated by the eight vectors e_i and $-e_i + 2 \sum_{j \neq i} e_j$, then $\sigma^\vee$ has twelve minimal generators $2e_i^* + e_j^*$, $i \neq j$.

8. See [Shaf, §V.1], [Hart, Ch. I].

9. If this is difficult now, try it after you have read about orbits in Chapter 3.

10. The quotient field of $\mathbb{C}[S]$ is the same as that of $\mathbb{C}[S+(-S)]$, so one can assume S and S' are sublattices; choose a basis for S' so that multiples of these elements form a basis for S.

11. If J is the ideal generated by such relations, the fact that the canonical map from $\mathbb{C}[Y_1, \ldots, Y_t]/J$ to A_σ is an isomorphism follows from the fact that the χ^u for u in S_σ form a basis for A_σ, and the monomials mapping to a given χ^u differ by elements of J. Finding a minimal generating set for this ideal is more subtle, cf.

> B. Sturmfels, "Gröbner bases of toric varieties," Tohoku Math. J.
>   **43** (1991), 249-261.

12. Two of the many references are

J. Herzog, "Generators and relations of abelian semigroups and semigroup rings," Manuscripta Math. **3** (1970), 175-193.

M. Hochster, "Rings of invariants of tori, Cohen-Macaulay rings generated by monomials and polytopes," Ann. Math. **96** (1972), 318-337.

Many of the results about toric varieties extend to such rings, and to the "nonnormal toric varieties" one gets by gluing their affine varieties together.

13. See

J. Herzog and E. Kunz, "Die Wertehalbgruppe eines lokalen Rings der Dimension 1," S. B. Heidelberger Akad. Wiss. Math. Natur. Kl. (1971), 27-67.

14. See [Shaf, Ch. V, §4.3].

15. In §2.3 we will see how to recover Δ from $X(\Delta)$. For the general proof see [Oda', Theorem 4.1].

16. Both are equivalent to the cone over $K \times 1$ meeting $E \times 0$ only at the origin.

17. Given the eight real numbers r_v, choosing three generators of each σ determines the vector u_σ, and the equation $\langle u_\sigma, v \rangle = r_v$ for the fourth generator is then a linear equation among the eight numbers. These six equations have a solution space of dimension three.

Chapter 2

1. A point given by a homomorphism $x: S_\sigma \to \mathbb{C}$ of semigroups is fixed exactly when $x(u) = 0$ for all $u \neq 0$, since if $u \neq 0$ there is a point in the torus $t: M \to \mathbb{C}^*$ with $t(u) \neq 1$, and $(t \cdot x)(u) = t(u)x(u)$. For σ not to span $N_\mathbb{R}$ means that S_σ contains some nonzero u together with $-u$, so $x(u)x(-u) = 1$.

2. See [Oda, §3.2] for Ishida's more intrinsic and more general construction of these complexes.

3. Use the simultaneous diagonalizability of commuting matrices; see §15 of

J. Humphreys, *Linear Algebraic Groups*, Springer-Verlag, 1975.

4. Diagonalizing, one may assume $T = (\mathbb{C}^*)^r$ acts on $V = \mathbb{C}^n$ by

$$(z_1, \ldots, z_r) \cdot e_j = (z_1^{m_{1j}} \cdot \ldots \cdot z_r^{m_{rj}}) e_j .$$

The ring of invariants is $\mathbb{C}[S]$, where

$$S = \{(p_1, \ldots, p_n) \in (\mathbb{Z}_{\geq 0})^n : \Sigma m_{ij}p_j = 0 \; \forall i\} .$$

5. The reference is

> J. Gubeladze, "Anderson's conjecture and the maximal monoid class over which projective modules are free," Math. of the USSR - Sbornik **63** (1989), 165-180.

For more on this see the article by Swan mentioned in Note 6 to Chapter 1.

6. For more on quotient singularities, see

> F. Hirzebruch, "Über vierdimensionale Riemannsche Flächen mehrdeutiger analytischer Funktionen von zwei komplexen Veränderlichen," Math. Ann. **126** (1953), 1-22.

> D. Prill, "Local classification of quotients of complex manifolds by discontinuous groups," Duke Math. J. **34** (1967), 375-386.

> E. Brieskorn, "Rationale Singularitäten komplexer Flächen," Invent. Math. **4**, (1968), 336-358.

> O. Riemenschneider, "Deformationen von Quotientensingularitäten (nach zyklischen Gruppen)," Math. Ann. **209** (1974), 211-248.

> F. Ehlers, "Eine Klasse komplexer Mannigfaltigkeiten und die Auflösung einiger isolierter Singularitäten," Math. Ann. **218** (1975), 127-156.

7. (a) For U_{σ_i}, the identification with the preceding exercise is by $m = d_i$ and $(a_1, \ldots, a_n) = (d_0, \ldots, d_{i-1}, d_{i+1}, \ldots, d_n)$. (b) Take N' generated by the v_i. This realizes $\mathbb{P}(d_0, \ldots, d_n)$ as $\mathbb{P}^n/G$, where $G = N/N' \cong \mu_{d_0} \times \ldots \times \mu_{d_n}/\mu_c$, where c is the greatest common divisor of the d_i. For more on twisted projective spaces see

> S. Mori, "On a generalization of complete intersections," J. Math. Kyoto Univ. **15** (1975), 619-646.

8. If φ is an algebraic group homomorphism, and the corresponding map $\varphi^*: \mathbb{C}[T, T^{-1}] \to \mathbb{C}[T, T^{-1}]$ takes T to $F(T)$, show that $F(T_1 T_2) = F(T_1)F(T_2)$, and deduce that $F(T) = T^k$ for some k.

9. Such limits play an important role in invariant theory:

> D. Mumford and J. Fogarty, *Geometric Invariant Theory*, Springer-Verlag, 1982.

For constructions of quotients of toric varieties by subtori of the given torus, see

> M. M. Kapranov, B. Sturmfels, and A. V. Zelevinsky, "Quotients of

toric varieties," Math. Ann. **290** (1991), 643-655.

10. When f is not proper, and X is embedded in a variety $\overline{X}$ so that the morphism f extends to a proper morphism from $\overline{X}$ to Y — which is in fact always possible — it is easy to construct such a discrete valuation ring and maps. In fact, one can take R to be the ring $\mathbb{C}\{t\}$ or $\mathbb{C}[[t]]$ of convergent or formal power series, with the maps corresponding to an analytic map of a small disk, with center mapping to a point of the closure of $\overline{X}$ that is not in X. For a formal proof, see [Hart, Ch. II, Exer. 4.11].

11. Look at the covering by the affine open U_σ and consider the distinguished points x_τ. For more on this, see [Oda, §1.5].

12. Let N be the lattice of rank $r+n-1$ generated by vectors $w_1, \dots, w_r$ and $v_0, \dots, v_n$, with relations

$$w_1 + \dots + w_r = 0 , \quad v_0 + \dots + v_n = a_1 w_1 + \dots + a_r w_r .$$

Let Δ be fan consisting of cones generated by subsets of these vectors that do not contain all of the w_i's or all of the v_j's. See [Oda, §1.7] for more on toric bundles.

13. Given the descriptions of the points in terms of semigroup homomorphisms, this is equivalent to the fact that an element u in $\sigma^\vee$ is in $\sigma^\perp$ exactly when $\varphi^*(u)$ is in $(\sigma')^\perp$, which follows from the fact that $\varphi(\sigma')$ contains points in the relative interior of σ.

14. The simplest example has rays also through $(2,1,1)$, $(2,1,2)$, and $(3,1,2)$, with cones as indicated:

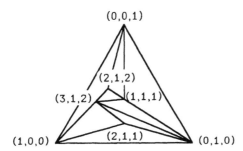

For more on this see

> M. Teicher, "On toroidal embeddings of 3-folds," Israel J. Math. **57** (1987), 49-67.

Recently it has been shown that any two nonsingular complete toric varieties of the same dimension can be obtained from each other by a sequence of maps that are blow-ups along smooth toric centers or the inverses of such maps. In fact, there are two independent proofs:

J. Wlodarczyk, "Decompositions of birational toric maps in blow-ups and blow-downs. A proof of the weak Oda conjecture," preprint.

R. Morelli, "The birational geometry of toric varieties," preprint.

15. Write $v_j = -av_i + bv_{i+1}$ and $v_{j+1} = -cv_i - dv_{i+1}$, with a, b, c, and d positive integers. The determinant of $\begin{pmatrix} -a & b \\ -c & -d \end{pmatrix}$ should be 1, but $ad + bc \geq 2$.

16. For (a), take the longest possible sequence of consecutive vectors in the same half-plane, and apply the topological constraint (see Note 15) to see where adjacent vectors can lie. For (b), write $v_j = -b_j v_0 + b_j' v_1$, and set $c_j = b_j + b_j'$; since $c_2 \geq 3$ and $c_i = 1$, there is a j with $c_j > c_{j+1}$ and $c_j \geq c_{j-1}$; show that $a_j = 1$.

17. The matrices give changes of bases from one pair of adjacent vectors to the next.

18. The condition (**) insures that the vectors go around the origin only once.

19. See the calculations of Note 3 to Chapter 1.

20. For example, see [Oda] and

V. E. Voskresenskii and A. A. Klyachko, "Toroidal Fano varieties and root systems," Math. USSR Izv. **24** (1985), 221-244.

V. Batyrev, "On the classification of smooth projective toric varieties," Tohoku Math. J. **43** (1991), 569-585.

21. For the relations among these generators — as well as the relation between the integers b_i and the integers a_j — see the article by Riemenschneider in Note 6.

22. The Hirzebruch-Jung continued fraction for $(k+1)/k$ produces a string of k 2's.

23. If $k \cdot k' \equiv 1 \pmod{m}$, write $k \cdot k' = 1 - s \cdot mb$; the matrix $\begin{pmatrix} k & m \\ s & -k' \end{pmatrix}$ maps $N = \mathbb{Z}^2$ isomorphically to itself and maps σ onto σ', giving an isomorphism between U_σ and $U_{\sigma'}$. The converse follows from the fact that the configuration obtained by the resolution process described here is the minimal resolution (none of the self-intersection numbers of exceptional rational curves is -1), which is uniquely determined by the singularity. This means that the singularity determines the sequence $a_1, \ldots, a_r$, up to replacing it by the reverse sequence $a_r, \ldots, a_1$, which corresponds to the pair (m,k') with $kk' \equiv 1 \pmod{m}$. Note that the singularity of U_σ is a cone over a lens space, and two such lens spaces are homeomorphic exactly when $m' = m$

and $k' = k$ or $kk' \equiv \pm 1 \pmod m$. For a reference, see the Hirzebruch article in Note 6.

24. The existence of such v is a simple case of a theorem of Minkowski, which can be seen by comparing the approximate number of points to the volume of large solids ($\Sigma t_i v_i$, $-m_i \le t_i \le m_i$). For the general case, see §24.1 of

> G. H. Hardy and E. M. Wright, *An Introduction to the Theory of Numbers,* fourth edition, Oxford Univ. Press, 1960.

25. The cone over a quadric was described in

> M. F. Atiyah, "On analytic surfaces with double points," Proc. Royal Soc. A **247** (1958), 237-244.

For flips, flops, and the relation to Mori's program see:

> J. Kollár, "The structure of algebraic threefolds — an introduction to Mori's program," Bull. Amer. Math. Soc. **17** (1987), 211-273.

> M. Reid, "Decomposition of toric morphisms," pp. 395-418 in *Arithmetic and Geometry II*, Progress in Math. **36**, Birkhäuser, 1983.

> M. Reid, "What is a flip?", preprint, 1992.

> T. Oda and H. S. Park, "Linear Gale transforms and Gelfand-Kapranov-Zelevinskij decompositions," Tohoku Math. J. **43** (1991), 375-399.

Recently, Batyrev has used toric varieties as a testing ground for the "mirror symmetry" conjectures coming from physics:

> V. V. Batyrev, "Dual polyhedra and mirror symmetry for Calabi-Yau hypersurfaces in toric varieties," preprint, 1992.

> V. V. Batyrev, "Variations of the mixed Hodge structures of affine hypersurfaces in algebraic tori," Duke Math. J. **69** (1993), 349-409.

Chapter 3

1. Any M-graded ideal (or submodule of the field of rational functions) has the form $\oplus \mathbb{C} \chi^u$, the sum over some subset of M; see Note 3 to Chapter 2. For complete proofs of these facts, including the following exercise, see [Oda', §5].

2. This follows from the fact that $\varphi_*(x_{\tau'}) = x_\tau$, because φ_* takes

$T_{N'}$-orbits to T_N-orbits. The last assertion uses the properness of φ_*.

3. Note that normality is crucial; the fact that the closed complement is small is not enough. The complement of a point in an irreducible variety can be simply connected without the variety being simply connected — for example if the variety is constructed by identifying two points in a simply connected variety. For an amplification of these points, see

> W. Fulton and R. Lazarsfeld, "Connectivity and its applications in algebraic geometry," Springer Lecture Notes **862** (1981), 26-92.

4. This is true with any coefficient group or sheaf: if $\mathcal{C}^{\cdot}$ is an injective resolution of the sheaf, this is the spectral sequence of the double complex $\bigoplus \mathcal{C}^q(U_{i_0} \cap \ldots \cap U_{i_p})$, with the vertical maps coming from the complex $\mathcal{C}^{\cdot}$ and the horizontal maps the "Čech" maps of alternating sums of restrictions. For details, see [Hirz, §I.2] and

> R. Godement, *Topologie algébrique et théorie des faisceaux*, Act. Sci. et Ind., Hermann, Paris, 1958.

5. See [Hirz, §I.4] for basic facts about Chern classes.

6. If the data $\{f_\alpha\}$ is used to define Cartier divisors, the data $\{f_\alpha g_\alpha\}$ defines the same Cartier divisor whenever g_α is a nowhere zero regular function on U_α; in addition, one identifies data with equivalent restrictions to a common refinement of the open covering. If the variety V meets the affine open set U_α, the ideal $\mathfrak{p}$ of $V \cap U_\alpha$ is a prime ideal in the affine ring A of U_α, and the local ring of V is the localization $A_\mathfrak{p}$. If X is normal and V has codimension one, $A_\mathfrak{p}$ is integrally closed and one-dimensional, so a discrete valuation ring. The order of D at V is the order of f_α with respect to this discrete valuation. That a Cartier divisor is determined by its Weil divisor follows from the fact that A is the intersection of these rings $A_\mathfrak{p}$. For more on this see [Shaf, §III.1], [Hart, §II.6], and [Fult, §2.2].

7. There is a $u \in S_\sigma$ such that $\langle u,v \rangle = 1$, where v is the first lattice point along τ.

8. This is local, so it is enough to do it on an affine U_σ, in which case it follows from the preceding lemma.

9. This exact sequence follows readily from the definitions, see [Fult, §1.8].

10. For example, $X = T_N$, with $n \geq 2$. All algebraic line bundles are trivial, but $H^2(T_N) = \wedge^2 M \neq 0$; the torus has analytic line bundles that are not algebraic.

11. We have seen that (i) $\Rightarrow$ (ii), and (ii) $\Rightarrow$ (iii) $\Rightarrow$ (iv) are clear from

the proposition. If σ is a maximal cone that is not simplicial, with generators $v_1, \ldots, v_r$, one can find positive integers a_i so that the points $(1/a_i)v_i$ are not in a hyperplane, so there is no $u(\sigma)$ with $\langle u(\sigma), v_i \rangle = -k a_i$ for $1 \le i \le r$ and any positive k.

12. The graph of this map is the resolution $X(\tilde{\Delta})$.

13. Find a strictly convex function ψ with $\psi(v_i) \in \mathbb{Z}$. For the second assertion, use the generators of semigroups found in §2.6. In fact, if D is an ample divisor on an n-dimensional toric variety, then $(n-1)D$ is always very ample. This is proved in

> G. Ewald and U. Wessels, "On the ampleness of invertible sheaves in complete projective toric varieties," Results in Math. **19** (1991), 275-278.

14. If the corresponding function is given by $u(\sigma)$ on the maximal cone σ, both are equivalent to the condition that $\langle u(\sigma), v_j \rangle > -a_j$ whenever $v_j \notin \sigma$.

15. Use all of the hyperplanes spanned by (n-1)-dimensional cones.

16. For more on the projectivity of nonsingular toric varieties see

> P. Kleinschmidt and B. Sturmfels, "Smooth toric varieties with small Picard number are projective," Topology **30** (1991), 289-299.

More examples like those in the text can be found in

> M. Eikelberg, "The Picard group of a compact toric variety," Results in Math. **22** (1992), 509-527.

17. If $P = (u_1, \ldots, u_r)$, the affine ring of U_σ is generated by variables $Y_1, \ldots, Y_r$ (with $Y_i = \chi^{(u_i \times 1)}$), and relations generated by $\prod Y_i^{a_i} - \prod Y_i^{b_i}$ if $\sum a_i u_i = \sum b_i u_i$ and $\sum a_i = \sum b_i$. This is the homogeneous coordinate ring of X_P in $\mathbb{P}^{r-1}$.

18. It suffices to look at the restriction of φ to the torus T_N, where the map is a map of algebraic tori $T_N \to (\mathbb{C}^*)^r / \mathbb{C}^*$ determined by a homomorphism of lattices from N to $\mathbb{Z}^r / \mathbb{Z} \cdot (1, \ldots, 1)$. The dimension of $\varphi(X)$ is the rank of the image of N, which is the dimension of P.

Chapter 4

1. For the topology of moment maps for more general actions of tori on algebraic varieties, see

> M. Goresky and R. MacPherson, "On the topology of algebraic torus

actions," Springer Lecture Notes **1271** (1987), 73-90.

2. For more on moment maps, including generalizations of these statements and proofs, see [Jurk] and

> M. F. Atiyah, "Convexity and commuting Hamiltonians," Bull. London Math. Soc. **14** (1982), 1-15.

> M. F. Atiyah, "Angular momentum, convex polyhedra and algebraic geometry," Proc. Edinburgh Math. Soc. **26** (1983), 121-138.

> V. Guillemin and S. Sternberg, "Convexity properties of the moment mapping," Invent. Math. **67** (1982), 491-513.

> F. C. Kirwan, *Cohomology of Quotients in Symplectic and Algebraic Geometry*, Princeton Univ. Press, 1984.

> L. Ness, "A stratification of the null cone via the moment map," Amer. J. Math. **106** (1984), 1281-1325; appendix by D. Mumford, "Proof of the convexity theorem," 1326-1329.

3. This is a general construction for divisors with normal crossings. There is a spectral sequence $H^q(X, \Omega^p_X(\log D)) \Rightarrow H^{p+q}(X \smallsetminus D, \mathbb{C})$, with $\Omega^p_X(\log D) = \wedge^p(\Omega^1_X(\log D))$. Reference:

> P. Deligne, *Équations Différentielles à Points Singuliers Réguliers*, Springer Lecture Notes **163**, 1970.

4. For details and generalizations to singular toric varieties, see [Oda, Ch. 3].

5. Construct this inductively over the simplices of S. Note that, on S, Z is defined by the equation $\psi \leq u$ and Z' by the equation $u + 1 \leq \psi$, so there is a band between them on each simplex.

6. For the four isomorphisms use: (i) the first exercise; (ii) duality of cohomology and homology; (iii) the second exercise; (iv) the first exercise.

7. The sections of $\mathcal{O}_X(-\Sigma D_i)$ are computed as before. For $i > 0$, $u \in M$, setting $Z = \{v \in |\Delta| : u(v) \geq 0\}$, $Z' = \{v \in |\Delta| : u(v) + \kappa(v) \leq 0\}$, and $S = \{v \in |\Delta| : \kappa(v) = 1\}$, the same argument shows that

$$H^i(X, \Omega^n_X)_{-u} \cong H^i_Z(|\Delta|) \cong \tilde{H}^{i-1}(S \smallsetminus Z' \cap S) \cong \tilde{H}^{i-1}(Z \cap S),$$

which vanishes since $Z \cap S$ is the intersection of a strongly convex cone with a polyhedral sphere.

8. For more on the dualizing complex on toric varieties, see [Oda, §3.2].

9. One reason for this is the general Riemann-Roch theorem, which

gives an interpretation for the coefficients of the polynomial. We will discuss this in Chapter 5.

10. For this and generalizations to several divisors, see

> A. G. Khovanskii, "Newton polyhedra and the genus of complete intersections," Funct. Anal. Appl. **12** (1978), 38-46.

11. Serre realized in the 1960's, based on the Weil conjectures, that there should be such virtual betti numbers. This motivated the search for weights, mixed Hodge structures, and motives. See pp. 185-191 of

> A. Grothendieck, "Récoltes et Semailles: réflexions et témoignage sur un passé de mathématicien," Montpellier, 1985.

For the construction of the mixed Hodge structures, see

> P. Deligne, "Théorie de Hodge, II, III" Publ. Math. I.H.E.S. **40** (1971), 5-57, **44** (1974), 5-77.

Although these papers do not explicitly do the construction for cohomology with compact supports, the methods required to do this are similar. This has been carried out, on the dual Borel-Moore homology groups, in

> U. Jannsen, "Deligne homology, Hodge-D-conjecture, and motives," in *Beilinson's Conjectures on Special values of L-Functions*, (M. Rapoport, N. Schappacher, and P. Schneider, eds.), pp. 305-372, Academic Press, 1988.

These virtual polynomials are discussed in

> V. I. Danilov and A. G. Khovanskii, "Newton polyhedra and an algorithm for computing Hodge-Deligne numbers," Math. USSR Izv. **29** (1987), 279-298.

> A. H. Durfee, "Algebraic varieties which are a disjoint union of subvarieties," Lecture Notes in Pure and Appl. Math. **105** (1987), 99-102.

12. This is proved in [Dani, §14], based on

> J. H. M. Steenbrink, "Mixed Hodge structure on the vanishing cohomology," in *Real and Complex Singularities, Oslo, 1976*, (P. Holm, ed.), pp. 565-678, Sitjthoff & Noordhoff, 1977.

It can also be deduced from the fact that the intersection homology always has a pure Hodge structure, as in the discussion in §5.2.

13. Here is one way to prove this. Note that the equality $\chi(X) = \chi_c(X)$ is true whenever X is an even-dimensional oriented manifold, since $H_c^i(X)$ and $H^{dim(X)-i}(X)$ are dual vector spaces. For a general X, take a covering by a finite number of affine open sets X_α; Mayer-

Vietoris sequences imply that

$$\chi(X) = \sum (-1)^{r+1} \chi(X_{\alpha_1} \cap \ldots \cap X_{\alpha_r}) \, ;$$
$$\chi_c(X) = \sum (-1)^{r+1} \chi_c(X_{\alpha_1} \cap \ldots \cap X_{\alpha_r}) \, .$$

It therefore suffices to show that $\chi(X) = \chi_c(X)$ if X is affine. Let
Y be the singular locus of X, and $U = X \smallsetminus Y$. It suffices by induction
on the dimension to show that $\chi(X) = \chi(Y) + \chi(U)$. Let $\pi: \tilde{X} \to X$ be
a resolution of singularities, with $\tilde{Y} = \pi^{-1}(Y)$ a divisor with (strong)
normal crossings. It follows by induction on the number of components
of $\tilde{Y}$ that $\chi(\tilde{Y}) = \chi_c(\tilde{Y})$, so $\chi(\tilde{X}) = \chi(\tilde{Y}) + \chi(U)$. There is a
neighborhood N of Y in X such that Y is a deformation retract of
N and $\tilde{Y}$ is a deformation retract of $\pi^{-1}(N)$. (For example, embed
X as a closed subvariety of $\mathbb{C}^m$ so that $Y = M \cap X$ for a linear
subspace M of $\mathbb{C}^m$, and take N to be the intersection of X with an
ε-neighborhood of M; the fact that π is proper implies that $\pi^{-1}Y$ is
a deformation retract of $\pi^{-1}(N)$.) By Mayer-Vietoris, the equation
$\chi(\tilde{X}) = \chi(\tilde{Y}) + \chi(U)$ is equivalent to the equation $\chi(\pi^{-1}(N) \smallsetminus \tilde{Y}) = 0$.
Since $\pi^{-1}(N) \smallsetminus \tilde{Y} = N \smallsetminus Y$, it follows that $\chi(N \smallsetminus Y) = 0$, and this is
equivalent to the equation $\chi(X) = \chi(Y) + \chi(U)$.

When Y is a point and N is a small neighborhood of Y, the
equation $\chi(N \smallsetminus Y) = 0$ says that the link of the singularity has zero
Euler characteristic. This was noticed by Sullivan:

D. Sullivan, "Combinatorial invariants of analytic spaces,"
    Springer Lecture Notes **192** (1971), 165-168.

Sullivan has shown that stratified spaces with odd-dimensional strata
have vanishing Euler characteristic. S. Weinberger, and M. Goresky
and R. MacPherson have verified that this can be used to extend the
results of this exercise to arbitrary spaces that can be stratified with
even-dimensional strata. For relevant techniques, see

M. Goresky and R. MacPherson, *Stratified Morse Theory*,
    Springer-Verlag, 1988.

Chapter 5

1. For general results about intersection theory, we refer to [Fult]. The
facts used in the proof of the proposition are proved in [Fult, §1.8 - 1.9].
Recently the author, R. MacPherson, and B. Sturmfels have shown
that the relations among the generators $[V(\sigma)]$ for $A_k(X)$ are
generated by those of the form $[\mathrm{div}(\chi^u)]$ for u in $M(\tau)$, and τ a
cone of Δ of dimension $n - k - 1$.

2. Assume that σ has independent (minimal) generators $w_1, \ldots, w_r$, and γ has one more generator $v = v_i$, and let $\tilde{e}$ be a lift of e to N_γ, so $N_\gamma = N_\sigma \oplus \mathbb{Z} \cdot \tilde{e}$. Take a positive integer integer p so that

$$p \cdot v = m_1 \cdot w_1 + \ldots + m_r \cdot w_r + p \cdot s \cdot \tilde{e},$$

for some integers m_j and $s = s_i$. Then

$$p \cdot \text{mult}(\gamma) = [N_\gamma : \Sigma \, \mathbb{Z} \cdot w_j + \mathbb{Z} \cdot p \cdot v]$$

$$= [N_\sigma : \Sigma \, \mathbb{Z} \cdot w_j] \cdot [\mathbb{Z} \cdot \tilde{e} : \mathbb{Z} \cdot p \cdot s \cdot \tilde{e}] = \text{mult}(\sigma) \cdot p \cdot s,$$

which shows that $s = \text{mult}(\gamma)/\text{mult}(\sigma)$.

3. For basic facts about intersecting with divisors, in particular the fact that rationally equivalent divisors on X determine rationally equivalent cycles on V, see [Fult, Ch. 2].

4. All the extremal rays in Mori's sense are determined by curves $V(\sigma)$, and one can see directly whether $V(\sigma)$ is numerically effective or not. See [Oda, §2.5] and the article by Reid in Note 25 of Chapter 2 for toric constructions of the corresponding contractions of such curves.

5. In general, when X is nonsingular, $V \cdot W$ can be constructed as a rational equivalence class of the expected dimension on $V \cap W$. For the construction of intersections on a nonsingular variety, see [Fult, Ch. 8].

6. When X is globally a quotient of a manifold M by a finite group G, then $A^*(X)_\mathbb{Q}$ can be identified with the ring of invariants in $A^*(M)_\mathbb{Q}$. When X is only locally such a quotient, the construction is harder, but there are now several ways to extend intersection theory to these varieties (and beyond):

> H. Gillet, "Intersection theory on algebraic stacks and Q-varieties," J. of Pure and Appl. Algebra **34** (1984), 193-240.

> A. Vistoli, "Alexander duality in intersection theory," Compositio Math. **70** (1989), 199-225.

> A. Vistoli, "Intersection theory on algebraic stacks and on the moduli spaces," Invent. Math. **97** (1989), 613-670.

> S. Kimura, "On varieties whose Chow groups have intersection products with Q-coefficients," University of Chicago thesis, 1990.

7. For this we follow [Dani], based on the article by Ehlers in Note 6 of Chapter 2, and

> J. Jurkiewicz, "Chow rings of projective non-singular torus embedding," Colloquium Math. **43** (1980), 261-270.

8. This problem is related to the "shellability" problem for cones. See

> A. Bjoerner, M. LasVergnas, B. Sturmfels, N. White, and G. Ziegler,

Oriented Matroids, Cambridge Univ. Press, 1993.

9. To see that $H^m(X) = IH^m(X) = H_{\dim(X)-m}(X)$ when X is an n-dimensional V-manifold, the point is that intersection homology is calculated by putting local conditions on cycles. There are now quite a number of surveys about intersection homology. The original paper is

M. Goresky and R. MacPherson, "Intersection homology theory," Topology **19** (1983), 135-162.

A recent survey, also emphasizing geometric intuition, is

R. MacPherson, *Intersection Homology and Perverse Sheaves,* AMS Lecture Notes, 1991.

For Hodge theory and intersection homology, see

M. Saito, "Mixed Hodge modules," Publ. Res. Inst. Math. Sci. (Kyoto Univ.) **26** (1990), 221-333.

10. For the relations between intersection homology betti numbers and the numbers of cones, including statements, proof, and some history, see

K.-H. Fieseler, "Rational intersection cohomology of projective toric varieties," J. Reine Angew. Math. **413** (1991), 88-98.

For more on these questions, see Note 35.

11. Let $\tilde{X} \to X$ be the resolution of singularities considered in §2.6, and let $\tilde{Y}$ be the union of the six lines in $\tilde{X}$ that is mapped to the set Y of six singular points in X. The facts that $H_i(\tilde{X},\tilde{Y}) = H_i(X,Y) = H_i(X)$ for $i \geq 2$, and that $\tilde{X}$ is the blow-up of $\mathbb{P}^3$ at four points give most of the answer, together with an exact sequence

$$0 \to H_3(X) \to H_2(\tilde{Y}) \to H_2(\tilde{X}) \to H_2(X) \to 0 .$$

To finish, the 6 by 5 matrix in the middle is calculated. Note in particular that $A_*(X) \to H_*(X)$ need not be surjective; replacing X by $X \times X$, one sees that the cycle map need not be surjective in even degrees. For a recipe for calculating the rational homology of a three-dimensional complete toric variety, see

M. McConnell, "The rational homology of toric varieties is not a combinatorial invariant," Proc. Amer. Math. Soc. **105** (1989), 986-991.

12. See [Dani, §10].

13. For Chern classes, Chern character, and Todd class in topology, see [Hirz]. For Chow theory, see [Fult, Ch. 3].

14. For the Hirzebruch-Riemann-Roch formula, as well as a discussion

of Grothendieck's extension to morphisms between smooth projective varieties, see [Hirz], and [Fult, Ch. 15]. For the extension to the singular case of Baum-Fulton-MacPherson, as well as the extension to the non-projective case by Fulton-Gillet, see [Fult, Ch. 18].

15. In the nonsingular case, $td(T_X) = 1 + \frac{1}{2}\Sigma D_i + \dots$. In the singular case, take a toric resolution of singularities, and note that divisors corresponding to any new edges are mapped to zero. Note in particular that $Td_{n-1}(X)$ need not be in the image of $Pic(X)$, since we have seen that ΣD_i need not be a $\mathbb{Q}$-Cartier divisor. The expression for $Td_0(X)$ follows from the fact that $\chi(X, \mathcal{O}_X) = 1$ when X is complete.

16. In the nonsingular case this can be found in [Dani, §11], based on

A. G. Khovanskii, "Newton polyhedra and toric varieties," Funct. Anal. Appl. **11** (1977), 289-296.

17. Morelli has given general formulas for the coefficients, showing that it *is* possible to assign the numbers r_σ in a way that depends only on σ:

R. Morelli, "Pick's theorem and the Todd class of a toric variety," Advances in Math., to appear.

Pommersheim has given some explicit formulas, relating some of these coefficients to Dedekind sums:

J. E. Pommersheim, "Toric varieties, lattice points and Dedekind sums," Math. Ann. **296** (1993), 1-24.

Generalizations have recently been announced by S. Cappell and J. Shaneson. For an approach using equivariant K-theory and Lefschetz-Riemann-Roch, see

M. Brion, "Points entiers dans les polyèdres convexes," Ann. sci. E. N. S. **21** (1988), 653-663.

18. A proof has recently appeared:

G. Barthel, J.-P. Brasselet, and K.-H. Fieseler, "Classes de Chern des variétés toriques singulières," C. R. Acad. Sci. Paris **315** (1992), 187-192.

19. For a combinatorial description of these numbers and a proof, see

P. Diaconis and W. Fulton, "A growth model, a game, an algebra, Lagrange inversion, and characteristic classes," Rend. Sem. Mat. Torino, in press.

Another proof has been given by Morelli in the paper of Note 17.

20. The coefficients r_σ can be taken to be 1 for $\dim(\sigma) = 0$, 1/2 for $\dim(\sigma) = 1$, 7/36 for $\dim(\sigma) = 2$, and 1/6 for $\dim(\sigma) = 3$. This

gives the formula $\#(v \cdot P) = 1 + (7/3)v + 2v^2 + (2/3)v^3$.

21. This follows from the fact that the intersection number can be realized by counting the number of intersections of inverse images of generic hyperplanes in the corresponding projective spaces. For a stronger result, proved jointly with R. Lazarsfeld, see [Fult, §12.2].

22. A modern treatment of the Bertini theorems is given in

> J.-P. Jouanolou, *Théorèmes de Bertini et Applications,*
> Birkhäuser Boston, 1983.

23. For a proof and discussion, see [Fult, Ex. 5.2.4]. Note that there is no need to take X to be nonsingular; the surface Y will then only be irreducible and complete, but the inequality $(D_1 \cdot D_2)^2 \geq (D_1 \cdot D_1)(D_2 \cdot D_2)$ is still valid for arbitrary Cartier divisors D_1 and D_2, provided at least one has nonnegative self-intersection; for example, one can apply the usual index theorem on a resolution of singularities of Y.

24. The inequalities follow formally from the case $k = 2$. For example, for $k = n = 3$,

$$V(P_1,P_2,P_3)^6 = V(P_1,P_2,P_3)^2 \cdot V(P_2,P_3,P_1)^2 \cdot V(P_3,P_1,P_2)^2$$

$$\geq \prod_{i \neq j} V(P_i,P_i,P_j) \ .$$

Taking two of these equal, one sees that

$$V(P_1,P_1,P_2)^3 \geq V(P_1,P_1,P_1)^2 \cdot V(P_2,P_2,P_2) = \mathrm{Vol}(P_1)^2 \cdot \mathrm{Vol}(P_2) \ ,$$

which is (b). This gives

$$V(P_1,P_2,P_3)^9 \geq \mathrm{Vol}(P_1)^2 \cdot \mathrm{Vol}(P_2) \cdot \mathrm{Vol}(P_2)^2 \cdot \mathrm{Vol}(P_3) \cdot \mathrm{Vol}(P_3)^2 \cdot \mathrm{Vol}(P_1) \ ,$$

and taking cube roots gives (c).

25. This can be deduced from the case $n = 2$, by the same formal manipulations as in the preceding note. For a direct proof for ample divisors, see §5 in

> J.-P. Demailly, "A numerical criterion for very ample line
> bundles," preprint.

The general case can be deduced from this by replacing E_i by $E_i + \varepsilon H_i$ with H_i ample, and letting ε go to zero — which is analogous to approximating bounded convex sets by n-dimensional convex polytopes. (Thanks to L. Ein for pointing this out.)

26. Use (6) with the preceding exercise.

27. For more on mixed volumes, including references, history, and many more inequalities that can be deduced from these formulas, see [BZ, Ch. 4].

28. New results on Newton polyhedra have been found in a series of papers by Gelfand, Kapranov, and Zelevinsky; see, e.g.,

> I. M. Gelfand, M. M. Kapranov, and A. V. Zelevinsky, "Newton polytopes of the classical resultant and discriminant," Advances in Math. **84** (1990), 237-254.

29. See

> D. N. Bernstein, "The number of roots of a system of equations," Funct. Anal. Appl. **9** (1975), 183-185,

> A. G. Kouchnirenko, "Polyèdres de Newton et nombres de Milnor," Invent. Math. **32** (1976), 1-31.

30. This is a result of the author and R. Lazarsfeld, the point being that any non-isolated components must give nonnegative contributions to the total intersection number. For a proof, see [Fult, §12.2].
For a generalization in another direction — to other groups — see

> B. Ya. Kazarnovskii, "Newton polyhedra and the Bezout formula for matrix-valued functions of finite-dimensional representations," Funct. Anal. Appl. **21** (1987), 319-321.

31. The references are

> L. J. Billera and C. W. Lee, "A proof of the sufficiency of McMullen's condition for f-vectors of simplicial convex polytopes," J. Combin. Theory (A) **31** (1981), 237-255.

> R. Stanley, "The number of faces of a simplicial convex polytope," Advances in Math. **35** (1980), 236-238.

32. For a proof of Macaulay's result, see

> G. F. Clements and B. Lindström, "A generalization of a combinatorial theorem of Macaulay," J. Combin. Th. **7** (1969), 230-238.

33. For a discussion of these problems, with references, see

> R. Stanley, "The number of faces of simplicial polytopes and spheres," in *Discrete Geometry and Convexity* (J. Goodman et al., eds.), Ann. New York Acad. Sci. **440** (1985), 212-223.

> C. W. Lee, "Some recent results on convex polytopes," Contemporary Math. **114** (1990), 3-19.

34. The dual of a simple polytope is simplicial.

35. For more on the relations between toric varieties and face numbers, see

> R. Stanley, "Generalized *h*-vectors, intersection cohomology of toric varieties, and related results," in *Commutative*

Algebra and Combinatorics, pp. 187-213, Advanced Studies
    in Pure Math. **11**, 1987.

R. Stanley, "Subdivisions and local *h*-vectors," J. Amer. Math.
    Soc. **5** (1992), 805-851.

This intrusion of algebraic geometry into the world of polytopes has
provided a challenge to combinatorialists. Recently G. Kalai and P.
McMullen have announced proofs of Stanley's theorem that do not
depend on toric varieties. McMullen's recent preprint, "On simple
polytopes," provides a challenge back to algebraic geometers: to
understand better intersection theory on singular toric varieties.

REFERENCES

[AMRT]   A. Ash, D. Mumford, M. Rapoport, and Y. Tai, *Smooth Compactification of Locally Symmetric Varieties,* Math. Sci. Press, 1975.

[Beau]   A. Beauville, *Complex Algebraic Surfaces,* London Math. Soc., 1983.

[Bryl]   J.-L. Brylinski, "Eventails et variétés toriques," in *Séminaire sur les singularités des surfaces,* Springer Lecture Notes **777** (1980), 247-288.

[BZ]   Y. D. Burago and V. A. Zalgaller, *Geometric Inequalities,* Springer-Verlag, 1988.

[Dani]   V. Danilov, "The geometry of toric varieties," Russ. Math. Surveys **33** (1978), 97-154.

[Dema]   M. Demazure, "Sous-groupes algébriques de rang maximum du groupe de Cremona," Ann. Sci. Ecole Norm. Sup. **3** (1970), 507-588.

[Fult]   W. Fulton, *Intersection Theory,* Springer-Verlag, 1984.

[Hart]   R. Hartshorne, *Algebraic Geometry,* Springer-Verlag, 1977.

[Hirz]   F. Hirzebruch, *Topological Methods in Algebraic Geometry,* 3rd ed., Springer-Verlag, 1966.

[Jurk]   J. Jurkiewicz, "Torus embeddings, polyhedra, k^*-actions and homology," Dissertationes mathematicae CCXXXVI, Polska Akademia Nauk, Instytut Matematyczny, Warszawa, 1985.

[KKMS]   G. Kempf, F. Knudsen, D. Mumford, and B. Saint-Donat, *Toroidal Embeddings I,* Springer Lecture Notes **339**, 1973.

[Oda]   T. Oda, *Convex Bodies and Algebraic Geometry,* Springer-Verlag, 1988.

[Oda']   T. Oda, *Lectures on Torus Embeddings and Applications,* TATA, 1978.

[Shaf]   I. R. Shafarevich, *Basic Algebraic Geometry,* Springer-Verlag, 1977.

[Teis]   B. Teissier, "Variétés toriques et polytopes," Séminaire Bourbaki no. 565, 1980/81, Springer Lecture Notes **901** (1981), 71-84.

INDEX OF NOTATION

INDEX